Studies in Logic
Volume 69

Logic and Conditional Probability
A Synthesis

Volume 58
Handbook of Mathematical Fuzzy Logic, Volume 3
Petr Cintula, Petr Hajek and Carles Noguera, eds.

Volume 59
The Psychology of Argument. Cognitive Approaches to Argumentation and Persuasion
Fabio Paglieri, Laura Bonelli and Silvia Felletti, eds

Volume 60
Absract Algebraic Logic. An Introductory Textbook
Josep Maria Font

Volume 61
Philosophical Applications of Modal Logic
Lloyd Humberstone

Volume 62
Argumentation and Reasoned Action. Proceedings of the 1st European Conference on Argumentation, Lisbon 2015. Volume I
Dima Mohammed and Marcin Lewiński, eds

Volume 63
Argumentation and Reasoned Action. Proceedings of the 1st European Conference on Argumentation, Lisbon 2015. Volume II
Dima Mohammed and Marcin Lewiński, eds

Volume 64
Logic of Questions in the Wild. Inferential Erotetic Logic in Information Seeking Dialogue Modelling
Paweł Łupkowski

Volume 65
Elementary Logic with Applications. A Procedural Perspective for Computer Scientists
D. M. Gabbay and O. T. Rodrigues

Volume 66
Logical Consequences. Theory and Applications: An Introduction.
Luis M. Augusto

Volume 67
Many-Valued Logics: A Mathematical and Computational Introduction
Luis M. Augusto

Volume 68
Argument Technologies: Theory, Analysis and Appplications
Floris Bex, Floriana Grasso, Nancy Green, Fabio Paglieri and Chris Reed, eds

Volume 69
Logic and Conditional Probability. A Synthesis
Philip Calabrese

Studies in Logic Series Editor
Dov Gabbay   dov.gabbay@kcl.ac.uk

Logic and Conditional Probability
A Synthesis

Philip Calabrese

© Individual author and College Publications 2017
All rights reserved.

ISBN 978-1-84890-258-9

College Publications
Scientific Director: Dov Gabbay
Managing Director: Jane Spurr

http://www.collegepublications.co.uk

Printed by Lightning Source, Milton Keynes, UK

All rights reserved. No part of this publication may be reproduced, stored in a retrieval system or transmitted in any form, or by any means, electronic, mechanical, photocopying, recording or otherwise without prior permission, in writing, from the publisher.

Preface

The main idea motivating this monograph is to represent (possibly uncertain) conditional information --- "A given B", "if B then A", A in the context of B", "A in case B", "whenever B then A" --- as ordered pairs (A|B) or (A/B) of propositions A and B. The condition B is the denominator of the fraction while the numerator A is whatever proposition is being identified given that condition. Since a proposition has a set m(p) of examples (models) in which it is realized (true), there is a corresponding collection of order pairs of examples (events) to which conditional probabilities may be assigned. The result is an algebra that successfully combines conditional logic with conditional probability while being completely faithful to both Boolean logic and standard Bayesian conditional probability theory.

This algebra implies a 3-valued logic of "true", "false" and "inapplicable" [Ch. 2] and extended operations of "AND", "OR" and "NOT" on order pairs of propositions, and even a reducible iterated conditioning operation ((A|B) | (C|D)) [See Ch. 2 & 3]. These extended operations of "AND" & "OR" exhibit so-called non-monotonic properties just as do corresponding natural language compound conditionals such as "(A if B) and (C if D)": Due to expanding the condition from B to (B OR D), the conditional proposition (A if B) can have smaller conditional probability than when conjoined with (C if D).

The same technique of ordered pairs is used in Chapter 4 to extend the operations of sum, difference, product and division for all numerical functions with possibly different or overlapping domains thus extending the standard operations for all random variables and all numerical functions. This allows a generalized mean value to be defined for functions with possibly overlapping domains.

Chapter 5 develops the formal algebraic structures that ensure that the model function m(p) can consistently carry probabilities (and conditional probabilities) from sets of models over to the propositions in which they are true. This chapter also includes examples of statements that are equivalent when wholly true but not equivalent when merely possible.

Chapter 6 demonstrates a remarkable variety of natural deductive relations in the context of conditionals. For instance, since the equation [(A|B) AND (C|D) = (A|B)] is not equivalent to [(A|B) OR (C|D) = (C|D)], they define different deductive relations in a hierarchy. These different deductive relations give rise to different deductively closed sets of propositions depending on which deductive relation is specified.

The ability to make Bayesian computations with this algebra has been well demonstrated in the SPIRIT program developed by W. Rödder [157, 125], which calculates maximum entropy solutions given partial information. [See Ch. 6]

Chapter 7 provides the connection to the fuzzy sets of L. Zadeh [58, 98] via "The probability that a proposition is necessary given a variable sets of axioms". Similarly, a proposition depending on time may go from merely possible to necessarily true after a time t.

Chapters 8 & 9 demonstrate how much of quantum logic is expressible using this system of conditional events including concepts of orthogonality, simultaneous verifiability, and completeness. While admitting that any hidden variable theory of objective positions and velocities of particles in space must admit faster than light interactions, this approach to quantum mechanics is completely consistent with the De Broglie–Bohm interpretation of quantum measurements.

Although a casual reader may want to skip over some of the subsection topics, they will likely be helpful to specialists looking to further develop this fertile, un-crowded research field.

Having been adapted from individual research papers, the chapters are fairly self-contained and can be read in any order.

Philip G. Calabrese
September 8, 2017

Table of Contents

Preface v
Table of Contents vii
Chapter 1. Introduction and Preview 1
  1.1 The Obstacle of Implication 1
  1.2 Material Implication versus Conditional Probability 2
  1.3 The Meaning of Partially True Propositions 4
  1.4 Algebraic Logic and Model Theory 5
  1.5 Strict Implication 6
  1.6 The Probability of a Proposition 7
  1.7 Additional Developments in Probability Logi 8
  1.8 Model Logic 8
  1.9 Multi-Valued Logics 9
  1.10 Fuzzy Sets 9
  1.11 Partially True Thinking 10
  1.12 An Example Calculation 10
  1.13 Formal Axioms of Boolean Logic and Probability 11
  1.14 Symbols and Conventions 13

Chapter 2. Conditional Events: Doing for Logic and Probability
    What Fractions Do for Integer Arithmetic 15
  2.1 Two-Valued Logic Versus Conditional Probability 15
  2.2 No Boolean Function for Conditional Probability 16
  2.3 The Idea of a Conditional Event Algebra 18
  2.4 Conditional Statements With or Without Certainty............ 22
  2.5 Development of Fractions for Integer Arithmetic 27
  2.6 Development of Fractions for Events and Propositions 28
  2.7 Truth Value Representation 32
  2.8 Partial Solution to an Expert System Circularity Problem 32
  2.9 A Problem of Iterated Conditioning 33
  2.10 Probabilities of Disjunctions & Conjunctions of Conditionals 34
  2.11 Overlapping surveillance regions 34
  2.12 Simpson's Reverse Paradox 35
  2.13 Cooperative Targeting Hypothesis Problem 38
  2.14 Independence and Conditionals 41
  2.15 Summary 42

Chapter 3. Logic and Probability Synthesized in Algebraic Logic and Model Theory 43
3.1 Truth Values Versus Probabilities 43
3.2 The Extension of a Proposition 44
3.3 "if - then - " in Logic Versus Probability 46
3.4 Formal Development of Conditional Probability Logic (CPL) ... 47
3.5 Conditional Events and Conditional Propositions 49
3.6 Definition of Equivalence for Conditional Propositions 50
3.7 Operations on Conditionals 51
3.8 Formulation of the Conditional Event Algebra (B|B) 53
3.9 Equivalent Conditionals 54
3.10 Operations on Conditionals 55
3.11 Operations on Conditionals - Further Motivations 57
3.12 Minimal Axioms for Operations in CPL 66
3.13 Theorem Characterizing CPL 66
3.14 Fundamental Theorem of Boolean Algebra for Conditionals ... 69

Chapter 4. Operating on Functions with Variable Domains 75
4.1 Synosis 75
4.2 Introduction 76
4.3 Extended Operations on Real-Valued Functions 77
4.4 Propositions, Events and Indicator Functions 78
4.5 Conditional Propositions, Events and Restricted Indicator Functions 79
4.6 Extended Operations on Conditional Propositions 80
4.7 Weighted Averages 83
4.8 Extended Operations on Random Variables 84
4.9 Conditional Random Variables 85
4.10 Operations on Conditional Random Variables 86
4.11 Expectations and Conditional Expectations 86
4.12 Lemma for Theorem 4.13 88
4.13 Theorem (Expected Value of a Sum of Conditional RandomVariables) 88
4.14 Theorem (Expected Value of a Product of Conditional Random Variables) 90
4.15 Definition of Independence of Random Variables 91
4.16 Corollary to Theorem 4.14 for Independent Variables 91
4.17 Work Force Example 91
4.18 Surveillance Region Example 91
4.19 Summary 93

Chapter 5. The Structure of Conditional Probability Logic 95
5.1 Ideals of Propositions 95
5.2 The Conditional Logic Generated by an Ideal 96
5.3 Logical and Set-Theoretical Operations on Ideals 98
5.4 Models of a Boolean Propositional Logic 100
5.5 Ideals of Models and Ideals of Conditional Events 101
5.6 The Extension Mapping 102
5.7 Probability of an Arbitrary Conditional Proposition 105
5.8 The Model Function m is an Isomorphism on Ideals 106
5.9 Statements Equivalent When Wholly True 106
5.10 The Probability of the Contrapositive of a Conditional.. 107
5.11 The Probability of the Converse 108
5.12 The System of Boolean Fractions....................................... 108
5.13 Formal Axioms of Conditional Probability Logic 109
5.14 Finite Conditional Probability Logics 110
5.15 Some Boolean Properties No Longer True 113
5.16 Standard Logical Formulas and Probabilities 113
5.17 Nonstandard Formulas & Associated Probabilities 114
5.18 Canonical Form of (q|p) ∨ (s|r) and (q|p) ∧ (s|r) 115
5.19 A Sample Calculation 117

Chapter 6. Deduction of Conditionals by Conditionals 119
6.0 Varieties of Deduction Premised on 3-Valued Propositions 119
6.1 Deductive Relations and Deductively Closed Sets of
    Conditionals 119
6.2 Extensions of Boolean Implication 121
6.3 Construction of Deductively Closed Sets of Conditionals 129
6.4 Generators of a Deductively Closed Set of Conditionals .. 131
6.5 The Exceptional Deductive Relations $\leq_{tr}$ and $\leq_V$ 143
6.6 Non-Elementary Examples of Deductively Closed Sets 147
6.7 Simplest Finite Deductively Closed Sets of Conditionals 158
6.8 Computations with Conditionals 171
6.9 Summary 175

Chapter 7. Fuzzy Sets, Time and Wholly-True Propositions ... 177
7.1 The Probability of Entailment --- The Fuzzy Connection 177
7.2 The Fuzzy Connection 179
7.3 Time 180
7.4 Logical Models as Instants of Time 180
7.5 Fuzzy Boundaries 181
7.6 Paradoxes of Self-Reference 182
7.7 Time and Quantum Logic 183
7.8 Time Revisited 185
7.9 Indicator Functions 186

Chapter 8. A More Natural Expression of Quantum Logic with Boolean Fractions 189
8.1 Introduction 189
 - Controversies of Quantum Logic 190
 - The case for Non-Local Realism 193
8.2 Boolean Fractions and Quantum Mechanics 196
 - The Conditional Event Algebras of Schay 197
 - Properties of Conditional Event Algebra 202
 - Theorem on Distributivity 202
 - Disjunction and Conjunction Superposition Formulas 207
 - Simultaneous Physical Measurements and Indeterminacy 209
 - Definition of Orthogonality for Conditionals 210
 - Theorem on Orthogonal Closure 211
8.3 Quantum Logic & Conditional Events 212
 - Definition of Simultaneous Verifiability 212
 - Theorem on Simultaneous Verifiability 213
 - Definition (Simultaneous Falsifiability) 216
 - Corollary on Simultaneous Verifiability and Falsifiability 216
 - Theorem on Boolean Sub-Algebras 216
 - Hilbert Space and Other Formulations 219
 - Theorem on Orthoalgebras 219
 - True, Wholly True and Given True 221

Chapter 9. The Logic of Quantum Measurements in terms of Conditional Events 225
9.0 Introduction 225
9.1 Principles of Quantum Mechanics.................................. 226
   - Formulation of Standard Algebra of Quantum Measurements 227
   - The Need for Explicit Conditions 227
   - Compatible Propositions and Boolean Sub-Algebras 228
9.2 Three-Valuedness of Conditionals................................... 230
   - Definition of Equivalence of Conditionals 230
   - Non-Monotonicity of Compound Conditionals 231
9.3 Simultaneous Verifiability of Conditionals 232
   - Orthogonal Expansion Theorem 233
   - Simultaneous Physical Measurements 235
   - Theorem on Generating Boolean Sub-Algebras 236
9.4 Orthogonality and Conditionals 237
   - Definition: Complete Set of Orthogonal Conditionals (COSC) 237
   - Completeness Characterization Theorem 238
   - Expansion Theorem 239
   - Representation Theorem 239
   - Completion Theorem 240
   - Restriction Theorem 240
   - Quantum Conditioning 241
   - Quantum Vectors and Linear Operator Language 242
   - Gleason's Theorem 244
   - Hidden Variables Versus Inherent Ambiguity 246
9.5 Deductive Logic and Quantum Operations 247
   - Deduction and Non-Monotonicity of Conditionals 247
   - The Sources of Non-Monotonicity 248
9.6 Bohm's Model of Quantum Mechanics 249
   - Heisenberg Indeterminacy 249
   - Quantum Entanglement 250
9.7 Conclusion 251

References and Bibliography 253

Index 267

Chapter 1
Introduction and Preview

> In the usual definition of conditional probabilities, the "probability of A given B" should be interpreted as "(probability of A) given B" rather than as "probability of (A given B)," since the notion of "A given B is not defined."
>
> G. Schay [60], "An Algebra of Conditional Events", 1968.

1.1 The Obstacle of Implication

One great obstacle that confronts anyone who would apply probabilities to arbitrary propositions is the differing formal treatments of the implication relation in the theory of logic[1] versus the theory of probability[2]. Both logic and probability are founded upon the common structure of a Boolean logic £ (see Section 1.13) which is equivalent[3] to the familiar algebra of subsets of some set under the operations of ∪ (union), ∩ (intersection), and ' (complement). In logic these operations refer to "and", "or", and "not", symbolized respectively by ∧ (conjunction), ∨ (disjunction), and ~ or ¬ (negation). In probability these operations retain their set-theory interpretation and notation. As for the objects being operated upon, in logic we speak about a set L of all propositions p, q, r..., while in probability we speak of a collection $\mathcal{B}$ of all events A, B, C.... This great commonality between logic and probability is based upon the fact that much of our logical and probabilistic thinking boils down to constructing sets and operating upon sets with Boolean operations. But the similarity disappears as soon as compound propositions (or events) of the form "if p then q" make their inevitable appearance.

---

[1] Logic was initially mathematized by G. Boole in his 1854 book [7]. Material implication has been developed by B. Russell [50], W. V. Quine [42, 102], and many others. See [5] for a lucid survey.

[2] As told by George Boole [7], P. de Fermat, B. Pascal, and C. de Mere initiated the theory of probability (1654). It was later axiomatized by A. N. Kolmogorov [27, 28]. More recently texts on probability have become ubiquitous.

[3] The famous 1936 representation theorem by M. H. Stone [54]

1.2 Material Implication versus Conditional Probability

In logical theory the all but standard definition of implication has become "material implication"[4]. "If p then q" is defined to be "q or not p", the so-called material conditional. In symbols this is "q ∨ ~p". It is also represented as "p ⊃ q". To say "if p then q is true" is to say "q ∨ ~p = 1", where " = " is the equivalence relation that defines equality in the Boolean logic.

By contrast, in probability theory prior to 1985 only a few researchers[5] in the foundations of probability had offered a definition of "if event A then event B". Although such a conditional event, "B given A", symbolized by (B|A), has usually been left formally undefined, there is a well-known standard definition for the conditional probability of event B given event A, symbolized by "P(B|A)". In probability theory P(B|A) is defined to be P(A ∩ B)/P(A), the ratio of the probability of the occurrence of both A and B to the probability of the occurrence of A. Note that P(B|A) is undefined if P(A) = 0. Now it has been observed Calabrese [9] (see Corollary 2.1.2 and Figure 1.1 below) that the probability of the material conditional, P(q ∨ ~p), is generally greater than P(q|p), the conditional probability of q given p, except when P(p) = 1 or P(q|p) = 1. Nor is there any other Boolean function f(q, p) which can be assigned P(q|p). (See Theorem 2.2.1.) Thus if P(q|p) is to be the truth measure of the conditional proposition "if p then q", then a new relation (q|p) "q given p" must be adjoined to the usual Boolean operations of ∧, ∨, and ~. [Here, (q|p) is equivalent to "q if p", and this will be distinguished from "q=1 if p" and its equivalent form "if p then q=1".]

---

[4] As propounded by B. Russell [50] in 1913.
[5] Among them are B. De Finetti [59], G. Schay [60], D. Scott and P. Krauss [51], Z. Domotor [15], A. Renyi [43], P. Suppes et al. [53,25], E. W. Adams [2]. T. Hailperin [20], D. Nute [36], and E. Lusk [32].

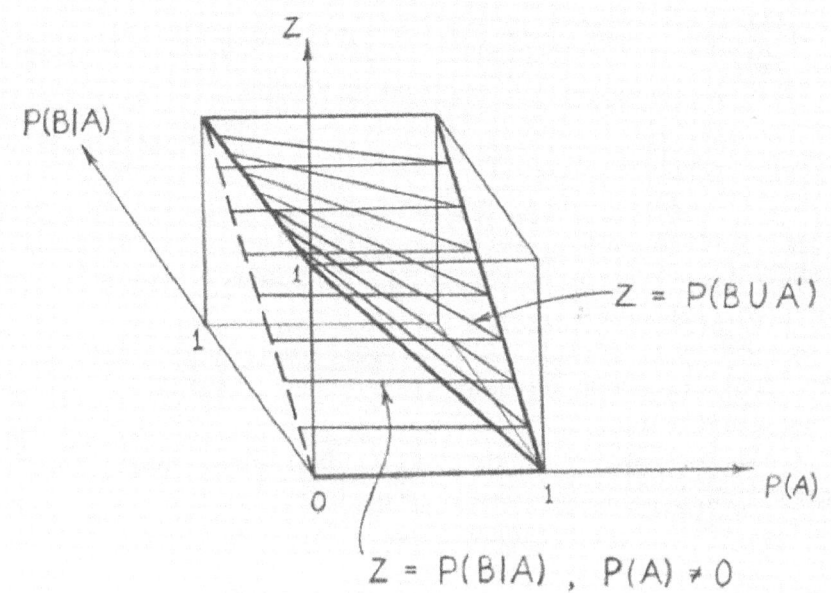

Figure 1.1 The Quantitative Difference between P(B|A) and P(B ∪ A′)

The differing treatment of implication in logic and probability is unfortunate— one might go so far as to say schizophrenic. For instance, there is the intuitionist phobia against two-valued proofs by contradiction. This aversion is justified when things are probable, but hardly in the two-valued logical thinking of mathematics.

Even so, in a logic with only two truth values—always true or always false, necessary or impossible, one or zero—there is still the well-recognized truth-table anomaly in which, according to material implication, "if p then q" is true whenever p is false. If p is false (the argument goes) then ~p is true, and so too is (q ∨ ~p) no matter what the truth-value of q. By contrast, in probability, if p is false, then the probability of p is zero and so the conditional probability ratio, P(p ∧ q)/P(p), is undefined. This is a telltale inadequacy of material implication that has been noticed by generations of introductory logic students: Does a scientist include cases in his sample for which the premise of his hypothesis is false? Does he count such cases as positive evidence for his hypothesis irrespective of the truth of the conclusion? No. The scientist does not report the probability that either the conclusion is true or the premise is false. He reports the conditional probability of the conclusion of his hypothesis given its premise, and so too must the logician who would consistently quantify the truth content of partially true statements.

Without a common axiomatic foundation for logic and probability embracing "if p then q", it has been impossible to formulate an efficient calculus wherewith to compute the probability of a complex logical proposition involving the operations of ∧, ∨, ~, and "if-then". Instead, the application of probabilities to logical constructions has remained an awkward exercise, the enigmatic ideal of specialists in probability instead of a universally used calculus for quantifying partial truth, a true algebra of logic and probability.

Nevertheless, while conditional probability theory has been hampered in its theoretical development by a lack of sufficient underlying axiomatic structure, it has still been successfully applied, for instance, in an area so difficult to imagine as quantum theory, thus showing its inherent utility and adaptability. In quantum theory it has sometimes been taken to be a physical law instead of an intellectual law[6].

1.3 The Meaning of Partially True Propositions

A second obstacle in the way of a synthesis of the foundations of logic and probability is the question of exactly how to attach real-number probabilities to logical propositions. What does it mean to say that a proposition is partially true? How can consistent truth-values (probabilities) be assigned to partially true propositions? What is the meaning of the probability of an arbitrary proposition?

Aristotle himself (354 B.C.), the father of syllogistic logic, realized that some statements such as "tomorrow there will be a sea battle" must be classified as "contingent", rather than simply true or false. A contingent statement is one that is not always true but rather is true in some cases (or contingencies) and false in others. In a formal axiomatic system of propositions such a statement may be adjoined to the axioms as a so-called independent axiom—one not provable from the others. Euclid's famous fifth postulate of geometry is such a statement, since it is true in some examples (models) of the four-postulate geometry and false in others. Thus it is that the study of partially true statements (or propositions) leads directly to the topic of related axio-

---

[6] Suppes [25] gives a lucid account of the belief in physically objective probability. He also provides necessary axioms for a "quantitative conditional probability structure." P. Calabrese [10] attaches probabilities to geometric distance and computes the probability of the triangle inequality.

matic systems and their associated models, to algebraic logic [33-35, 55] and model theory [14].

1.4 Algebraic Logic and Model Theory

In algebraic logic a central concept is that of an ideal J (filter, sum ideal) of propositions. An ideal is a set L of propositions that is closed under conjunction ($\wedge$) and arbitrary disjunction ($\vee$). (See Definition 5.1.1.) Ideals are also characterized by the property that they contain (p $\wedge$ q) if and only if they contain both p and q. The set T of all axioms and theorems (all propositions necessarily entailed by the axioms) forms an ideal, also called a deductively closed set of propositions. Conversely, any ideal of propositions is a candidate for being the set of all axioms and theorems of some Boolean logic. Adjunction to the axioms of a proposition that is not already necessary, nor impossible, results in a larger ideal T of always true propositions and establishes new equivalences among the remaining partially true propositions and ideals of L. The ideal generated by a set X of propositions, that is, the smallest ideal J that includes X, is called the *intension* of X. Adjoining a proposition that is inconsistent with the axioms results in a system in which all statements become true, and therefore also false, which means that the new system is worthless as a conceptual framework.

Along with a Boolean proposition logic there must also be a well-defined set of models, Ω, in which the initial axioms A are true. By definition [14, p. 4], in a model every proposition is either true or false. For example, a game of dice has axioms (rules) and a well-defined set of models (dice throws) that satisfy the rules. The subset of models m(J) in which the propositions of J are true is called the *extension* of J. Corresponding to the ideals of L there are the ideals of models. For any set J of propositions, its extension m(J) is an ideal of models (see Theorems 5.6.2 and 5.6.5).

In this context a logic is two-valued if and only if its ideal T of axioms and theorems is maximal, that is, an ultrafilter. That is, every proposition either is in T or has its negation in T. Every proposition is either true in all models or false in all models. No proposition that is not already always true can be adjoined as an axiom without making the logic inconsistent. For every proposition p, either p = 1 or p = 0. The probability of a proposition must be 0 or 1. Such are the characteristics of the two-valued logic.

The smallest ideal (p) containing a proposition p is also called the *intension* of p — all those propositions that are entailed by the axioms with the additional assumption of p. The *extension*, m(p), of p is the subset of all models of L in which p is true. For example, if in some geometric axiomatic system L, p is the proposition expressing the Pythagorean theorem, then the extension of p is the set of all models of L in which the Pythagorean theorem is true. The intension of p is the set of all propositions that necessarily follow in L from the truth of the Pythagorean theorem. "q in (p)" means that q is entailed by p and the axioms, "q in T" means that q is entailed by the axioms alone; that is, q is a theorem; that is, q is true in all models of the logic L.

Ideals can be combined by either logical or set-theoretic operations, with somewhat surprising results. For instance, if I and J are ideals, then $(I \vee J)$ is the same as $(I \cap J)$, but $(I \wedge J)$ is generally larger than $(I \cup J)$. (See Theorem 5.3.2.) Note also that propositions do filter the models in the sense that larger ideals of propositions have fewer models that satisfy them. (See Theorem 5.6.1 and Figure 3.1.)

1.5 Strict Implication

Along the same lines, and no doubt developed concurrently with algebraic logic, are the systems S1-S5 of Lewis and Langford [30]. In their absorbing theory of "strict implication" propositions are clearly portrayed as being of three kinds:

(1) Necessary—always true, true in all cases.
(2) Possible—sometimes true, true in some cases and false in others.
(3) Impossible—always false, false in all cases.

Thus "strict implication" is quite consistent with the theory of algebraic logic. Later it was shown[7] that p is necessary if and only if p = 1. Propositions that are true in some cases (models, examples) but false in other cases (models, examples) may be assigned a probability other than 0 or 1.

A more recent development in possibility theory has been offered by G. Shafer [52] and appears to include the formulation of probability logic presented here, but not vice versa. With proper interpretation the collection m(q), of all models in which a proposition q is true, is a belief (or credibility) function in the sense of Shafer. However, Shafer handles even mutually inconsistent data (a feature of everyday life)

---

[7] Paul Henle, Harvard University. See [30, p. 492].

using 'degrees of belief'. However, inconsistent propositions in the same logic always imply that in that logic all propositions are true.

1.6 The Probability of a Proposition

The probability, P(p), of a proposition p is defined[8] to be the probability of that (ideal) subset of models, m(p), in which p is true:

$$m(p) = \{\text{models in which p is true}\}, \qquad (1.6.a)$$

$$P(p) = P(m(p)). \qquad (1.6.b)$$

Since m(p) is the extension of p, P(p) is the probability of the extension of p.

For example, by having a suitable probability measure on some well-defined universe of models of the four-postulate geometry one could speak about the probability of Euclid's fifth postulate: the probability of those models of the four-postulate geometry in which the fifth postulate is also true. This illustrates just how an arbitrary proposition in an axiomatic context can be assigned a probability based on a measure of the models of the axiomatic system in which the proposition is true. It is easy to show that the mapping m is a homomorphism of the propositional ideals onto the model ideals for which ∧, ∨, ~ correspond to ∩, ∪, and ' respectively. That m is one-to-one is equivalent to K. Gödel's first-order completeness theorem [17] in a few steps. (See Theorem 5.6.5 and Figure 3.1.) For each ideal I of L there is also a natural homomorphism h of L onto a conditional logic L/I whose kernel (the inverse image of 1) is I.

These beautiful mathematical lines were initially developed by S. Mazurkiewicz [33-35] and A. Tarski [55, 56] and nicely expressed and elaborated by E. W. Beth [5] and others[9]. Carnap developed the definition of the probability of a proposition p. But without a resolution of obstacle number one concerning the implication relation, these promising explorations were suspended and soon were buried in the subsequent literature.

---

[8] G. Boole [7], R. Carnap and R. C. Jeffrey [13,37], H. Gaifman [16], D. Scott and P. Krauss [51], E. W. Adams [2], and T. Hailperin [20] all define the probability of a proposition as the probability of its extension set of models.

[9] Including G. Birkhoff [6], P. Rosenbloom [49], R. Carnap [11-13], P. Halmos [22-23], and C. C. Chang and H. J. Keisler [14].

1.7 Additional Developments in Probability Logic
In this book these lines are continued as follows. Given an ideal I, a conditional Boolean logic L/I is defined (see Definition 5.2.2) consisting of equivalence classes of propositions arising from the conditional adjunction of I to the axioms. This forms the basis for the definition of (q|p), the ideal generated by q in the conditional logic L/(p) formed by adding p to the axioms of L. "q|p" is a member of (q|p). Equivalent definitions apply to the models of L. The model function m can be extended to L/I and in particular to L/(p) by

$$m(q|p) = (m(q) | m(p)), \qquad (1.7.a)$$

so that (q|p) obtains the desired conditional probability, P(q|p), by the definition

$$P(q|p) = P(m(q|p)). \qquad (1.7.b)$$

Two conditional propositions (p|I) and (q|I) are equivalent (see Definition 5.1.2) if and only there is a proposition r in I such that $p \wedge r = q \wedge r$. (q|p) is equivalent to (s|p) if and only if $q \wedge p = s \wedge p$. It is also true that two propositions p and q are equivalent given an ideal I if and only if they generate the same ideal in L/I. (See Corollary 5.9.3.)

A definition of (q|p) is not very useful unless conditional propositions can be combined using $\wedge, \vee, \sim$, and | as in ordinary speech. See Definition 5.12 for the formal definition of the conditional closure L/L of a Boolean logic L. It turns out that the set *L/L* of conditional propositions defined by

$$L/L = \{(q|p): p, q \in L\} \qquad (1.7.c)$$

is closed under $\wedge, \vee, \sim$, and | according to now well established operations. (In Chapter 3)

1.8 Model Logic
Conditional Probability Logic (CPL) is a part of modal logic (See Rescher [48]). Among the modalities there are:

(1) The classical or alethic (possibility—p is possible, necessary).

(2) The epistemic (knowledge—p is known, believed, etc.)
(3) The temporal (time—p is true sometimes, all times, at one time, etc.)
(4) The boulomaic (desires—p is feared, hoped, etc.)
(5) The deontic (duties—p ought, should, etc.)
(6) The evaluative (values—p is good, bad, etc.)

Such a spectrum of logical relationships to explore!

1.9 Multi-Valued Logics

Other approaches to the problem of the assignment of more than two truth-values to propositions that deserve mention include the many versions of multivaled logic [38] initiated by the three-valued logic of J. Lukasiewicz [31]. Unfortunately, these logics follow the quite serviceable two-valued logic by defining, for instance, the truth-value of the compound proposition "p or q" wholly in terms of the truth values of p and of q. But the probability (truth value) of "p or q" depends also upon the probability of the proposition "p and q" according to the well-known formula

$$P(p \vee q) = P(p) + P(q) - P(p \wedge q). \tag{1.9.a}$$

1.10 Fuzzy Sets

Another, more promising approach to more general truth values is the fuzzy sets of L. Zadeh [58, 98]. A fuzzy set is one whose characteristic function $Z(x)$ takes values in the unit interval [0,1] instead of being restricted to the two-element set {0,1}. That is, given an element x and the fuzzy set A, there is some value $Z(x)$ that represents the degree to which x is an element of A.

The subsequent development of fuzzy sets has been extensive. See for instance Madan M. Gupta (Ed.) [19] or H. Prade [39] for a good survey and bibliography. B. Kosko [29] has offered fuzzy cognitive structures to deal with fuzzy and incomplete—even inconsistent — information. See I. R. Goodman & H. T. Nguyen [18] for a lucid account of the relationship between probability theory and fuzzy set theory and for the concept of a random subset. Random subsets subsume most of fuzzy set theory while maintaining a probabilistic approach. A special kind of fuzzy set is one whose characteristic function is a probability function. That is, there is some probability, $P(x \in A)$, that x is an element of A. This idea of a fuzzy set arises naturally in conditional probability logic when it comes time to consider the probability that a proposition is always true (true in all cases) as contrasted from the probability of the cases in which it is true. The former construction is $P(p \in T)$, the probability that p is in the ideal T, while the latter is $P(m(p))$, the probability of the subset of models in which p is true. (See Chapter 7.)

1.11 Partially True Thinking
Subtle issues are raised by the theory of probability logic. For instance, G. Boole realized that propositions that are equivalent in the two-valued logic might not be equivalent when considered only probable.

> "One remarkable circumstance which presents itself in such applications deserves to be specially noticed. It is, that propositions which, when true, are equivalent, are not necessarily equivalent when regarded only as probable.", G. Boole [7], *The Laws of Thought*, 1854

But his insight has apparently been underestimated even by a perceptive and sympathetic reviewer and developer of Boole's ideas (See [20] p 168).

For example, in the two-valued logic the contrapositive (~p | ~q) of a conditional proposition (q|p) is equivalent to it. Proof by contradiction is based upon this equivalence. But in probability logic (see Theorem 5.10.1 for a proof) the following equation holds as long as neither p nor ~p is impossible:

$$P(\sim p \mid \sim q) = P(q|p) + [1 - P(p)/P(\sim q)] [1 - P(q|p)] \qquad (1.11.a)$$

From this it necessarily follows that $P(\sim p \mid \sim q) = P(q|p)$ in case either of these conditional probabilities is one or in case the premises of the two conditionals have the same probability. So the contrapositive of a conditional proposition cannot in general be equivalent to the conditional proposition.

1.12 An Example Calculation
As engrossing as this abstract approach may be to the mathematical specialist, there comes a time to put down the tools of abstract algebra (homomorphisms, ideals, etc.), to remove the scaffolding, and to define a simple-to-use algebra of propositions and conditional proposi-

tions that everyone can use to assign and compute probabilities and conditional probabilities. The axioms of Boolean algebra (as given, for instance, in Section 1.13), together with Equations (1.6.a–b), (1.7.a–b) and the extended operations expressed in conditional event notation in subsection 2.6.1 or in conditional proposition notation in subsections 3.7.1-4, allow a straightforward reduction of any complex conditional expression into conditional propositions that have a common axiomatic condition. Probabilities can then be assigned to these propositions or perhaps to these in conjunction with other easily estimated conditional propositions in terms of which the original expression has been re-expressed.

In order to see how this new algebra can be applied to practical logical-probabilistic problems, a familiar problem should be solved by these new methods. How about a dice game? Consider the following proposition: "If you first role a 6, then you will win, or if you first roll a 4 you will lose." What is its probability? How true is it? See Section 5.19.

Finite conditional logics and their models can readily be implemented on a computer to facilitate the design of so-called artificially intelligent mechanisms that are better able to think in terms of absolutes in relation to partial (probabilistic) truths.

1.13 Formal Axioms of Boolean Logic and Probability

A Boolean algebra (logic) £ is a set of propositions L (including two constants 0 and 1) that is closed under the three operations $\wedge$ (conjunction), $\vee$ (disjunction), and $\sim$ (negation) and that satisfies the axioms:

$$p \wedge q = q \wedge p, \qquad\qquad p \vee q = q \vee p, \qquad\qquad (1.13.\text{a})$$
$$(p \wedge q) \wedge r = p \wedge (q \wedge r), \qquad (p \vee q) \vee r = p \vee (q \vee r), \qquad (1.13.\text{b})$$
$$1 \wedge p = p, \qquad\qquad 0 \vee p = p, \qquad\qquad (1.13.\text{c})$$
$$p \wedge (\sim p) = 0, \qquad\qquad p \vee (\sim p) = 1, \qquad\qquad (1.13.\text{d})$$
$$p \wedge (q \vee r) = (p \wedge q) \vee (p \wedge r), \quad p \vee (q \wedge r) = (p \vee q) \wedge (p \vee r), \quad (1.13.\text{e})$$
$$p \wedge p = p, \qquad\qquad p \vee p = p, \qquad\qquad (1.13.\text{f})$$

This is essentially the formulation as given by T. Hailperin [20].

With these axioms, the set L can be identified with the logic £, and sub-logics of £ are determined by the operationally closed subsets I of L.

" = " is an equivalence relation that may be added to the object language or remain in the syntax language. In the latter context, the "theorems" of £ can be described as those propositions p in *L* for which p = 1. For it is well known that two propositions p and q are equivalent if and only if both material conditionals are equivalent to 1. That is, p = q if and only if both q ∨ ~p =1 and p ∨ ~q = 1. Therefore every equivalence between propositions (including the axioms) can be expressed in terms of equivalences to 1. [There is also reason to believe that the familiar existential and universal quantifiers can be defined in terms of equivalence (=).]

A Boolean logic is also a lattice, usually constructed by defining the following partial ordering:

$$p \leq q \text{ if and only if } p \wedge q = p. \qquad (1.13 \text{ .g})$$

The least upper bound and the greatest lower bound are then p ∨ q and p ∧ q respectively. But from the point of view of ideals, it is natural to consider the dual lattice in which p ∧ q is the supremum and p ∨ q the infimum, in which 0 is the greatest and 1 the least element: for to assert both p and q is to entail at least the union of the entailments of p and q; while to assert p ∨ q is to entail only that which both p and q entail. The situation is reversed in the model realm, since p has at least as many models as has p ∧ q, while p ∨ q has no fewer than has p.

A Boolean function is one that can be formed by repeated (finite) use of the three Boolean operations (∧, ∨, and ~) upon a finite set of propositions p, q, r.... (Extension of Boolean algebra to countable conjunctions will not be considered here.)

Since both conjunction (∧) and disjunction (∨) are idempotent, and since double negation (~~) is also idempotent, a Boolean function of propositions p, q, r... can be expressed as a disjunctive polynomial whose terms are all conjunctions of those propositions or their negations.

By factoring out p from all terms having factor p in them and doing the same for terms having factor ~p, a Boolean function (i.e. polynomial) $f(p, q, r...)$ of a finite number of propositional variables p, q, r... can, by the "law of development", be expressed as follows:

$$f(p, q, r...) = [f(1, q, r...) \wedge p] \vee [f(0, q, r ...) \wedge (\sim p)].$$

By repeated applications of the law of development every such Boolean function can be canonically expressed as a disjunction of conjunctions of the propositional variables or their negations, each conjunction being conjoined with as associated value of f in which all variables are set equal to 1 or 0. It also follows that there are exactly 2^N distinct Boolean functions of N propositional variables, since there are that many different ways to assign a 1 or 0 to the variables of f.

A probability space $\mathcal{P}$, as defined by A. N, Kolmogorov [27, 28], is a universe Ω of possible occurrences, together with a collection $\mathcal{B}$ of events (subsets of possible occurrences closed under countable unions, intersections, and complements and including Ω), together with a function P (the probability measure) defined on $\mathcal{B}$ into the real interval [0,1] such that the following 3 axioms hold: Firstly, the probability of the set (event) of all possible occurrences is 1. That is,

$$P(\Omega) = 1 \tag{1.13.h}$$

Secondly, all events have non-negative probabilities. That is,

$$\text{For all events } B \in \mathcal{B}, \quad P(B) \geq 0 \tag{1.13.i}$$

Thirdly, if B_i for $i = 1, 2, 3\ldots$ are pair-wise disjoint events then

$$P(B_1 \cup B_2 \cup \ldots) = P(B_1) + P(B_2) + \ldots \tag{1.13.j}$$

In his book [27, 28] Kolmogorov proved that these axioms are so broad that they include all probabilities generated by self-consistent sets of (cumulative) distribution functions. In probability logic Ω is identified with a set of all models in which the propositions of some Boolean algebra are true. In this way an event becomes a subset of models.

1.14 Symbols and Conventions
Most of the symbols and conventions used here are sufficiently standard not to require special mention apart from the preceding introduction. A few, however, do deserve a comment. When two equivalence relations are being discussed, one will sometimes be denoted "≡" instead of "=". Nothing else is connoted by the use of ≡ versus =. In particular, "≡" does not mean "identical". The use of double quotes ("..."), as in "and" or "or", is meant to distinguish between object- and

syntax-language versions of the same concept. With respect to priority of operations when parentheses are omitted, negation (~) takes precedence and then conjunction (∧ or juxtaposition) followed by disjunction (∨) and then the conditional (|). Thus (r | qs ∨ ~p) means (r | ((qs) ∨ (~p))). A logic £ is usually identified with the set L of all of its propositions.

> Ordinary logic has two values and considers logical entities (propositions) capable of only two values: "true" or "false". Would that be insufficient? Is it necessary to consider a third modality (eg "possible"), or several others?
> Bruno de Finetti [90], The Logic of Probability (in French), 1936.

Chapter 2

Conditional Events: Doing for Logic and Probability What Fractions Do for Integer Arithmetic

> ... everything necessarily is or is not, and will be or will not be; but one cannot divide and say that one or the other is necessary. I mean, for example: it is necessary for there to be or not to be a sea battle tomorrow; but it is not necessary for a sea battle to take place tomorrow, nor for one not to take place—though it is necessary for one to take place or not to take place.
>
> Aristotle [181], in *De Interpretatione*

2.1 Two-Valued Logic Versus Conditional Probability

Two-valued logics and conditional probability spaces have both proved quite useful in the past, and any new theory must agree with both where they agree. Probability cannot get along without the conditional probability ratio $P(q \wedge p)/P(p)$ as the truth measure of $(q|p)$. On the other hand the truth-value of the material conditional $(q \vee \sim p)$ is certainly valid when p is true. The quantitative difference between these two can be worked out in a few steps. Let A and B be events in a standard probability space $\mathcal{P} = (\Omega, \mathcal{B}, P)$.

THEOREM 2.1.1 The probability of the material conditional, $P(B \cup A')$, is a quadratic function of $P(A)$ and $P(B|A)$, namely

$$P(B \cup A') = 1 - P(A) + P(B|A)P(A) \qquad (2.1.a)$$

except where $P(A) = 0$, in which case $P\{B \cup A'\} = 1$ and $P(B|A)$ is undefined.

Proof of Theorem 2.1.1 $(B \cup A') = A' \cup (B \cap A)$, where A' and $(B \cap A)$ are disjoint. Therefore $P(B \cup A') = P(A') + P(B \cap A) = 1 - P(A) + P(B|A)P(A)$ if $P(A)$ is not 0.

COROLLARY 2.1.2 $P(B \cup A') \geq P(B|A)$ with equality if and only if $P(A) = 1$ or $P(B|A) = 1$.

Proof of Corollary 2.1.2 $P(B \cup A') = 1 - P(A) + P(B|A)P(A) = 1 - P(A) - P(B|A) + P(B|A)P(A) + P(B|A) = [1 - P(A)][1 - P(B|A)] + P(B|A)$. Thus

$$P(B \cup A') = P(B|A) + [1 - P(A)][1 - P(B|A)] \quad (2.1.b)$$

Since the factors in the brackets are nonnegative, the left-hand side is greater than the right except when either of the factors is zero, that is, when either $P(A) = 1$ or $P(B|A) = 1$.

To bring home the quantitative difference between $P(B|A)$ and $P(B \cup A')$, equation (2.1.a) has been graphed in Figure 1.1. The curved surface is $P(B \cup A')$ as a function of $P(B|A)$ and $P(A)$. The plane $Z = P(B|A)$ is also easily identified (with a dashed line in case $P(A) = 0$, since $P(B|A)$ is then undefined). Note that the surface and the latter plane coincide on the exterior boundary where $P(A) = 1$ or $P(B|A) = 1$, but nowhere else.

As $P(A)$ and $P(B|A)$ approach 0, the difference between $P(B|A)$ and $P(B \cup A')$ approaches 1. Even if all events have either probability 0 or 1, we may still have $P(B \cup A') = 1$ while $P(A) = 0$ and $P(B|A)$ is undefined.

2.2 No Boolean Function for Conditional Probability

If the Boolean function $f: f(A, B) = B \cup A'$ (the material conditional) will not suffice, then perhaps one of the other 15 two-place functions[10] of the two-valued logic will work. But actually the situation is as stated in the following theorem.

---

[10] For an account of all 124 "Menger algebras" generated by 2-place composition of different subsets of these sixteen 2-place functions, see P. Calabrese [8].

THEOREM 2.2.1 There is no two-place Boolean function, f of events A and B for which

$$P(f(A, B)) = P(B|A) \qquad (2.2.a)$$

for all events A and B, $P(A) \neq 0$.

Proof of Theorem 2.2.1 By the standard definition of conditional probability, Equation (2.2.a) becomes (using juxtaposition for intersection $\cap$),

$$P(A) P(f(A, B)) = P(AB) \qquad (2.2.b)$$

Now every two-place Boolean function f can be expressed in the canonical form

$$f(A,B) = f(\Omega,\Omega)AB \cup f(\Omega,\phi)AB' \cup f(\phi,\Omega)A'B \cup f(\phi,\phi)A'B' \quad (2.2.c)$$

Substituting for $f(A,B)$ in Equation (2.2.b) and using the fact that the events in the brackets are disjoint, we have

$$P(A)[P(f(\Omega,\Omega)AB) + P(f(\Omega,\Omega)AB') \\ + P(f(\phi,\Omega)A'B) + P(f(\phi,\phi)A'B')] = P(AB) \qquad (2.2.d)$$

which must be true for all events A and B as long as A has positive probability. But if $A = B$, the Equation (2.2.d) reduces to

$$P(A) [P(f(\Omega, \Omega)A) + P(f(\phi, \phi)A')] = P(A) \qquad (2.2.e)$$

for all A with positive probability. It follows that both $f(\Omega, \Omega)$ and $f(\phi, \phi)$ must be equivalent to Ω. On the other hand, by setting $A = B'$ Equation (2.2.d) reduces to

$$P(A) [P(f(\Omega, \phi)A) + P(f(\phi, \Omega)A')] = 0$$

for all A with positive probability. Therefore, $f(\Omega, \phi)$ and $f(\phi, \Omega)$ must be equivalent to ϕ. Substituting these values for f into Equation (2.2.c), the only possible two-place function therefore is

$$f(A, B) = AB \cup A'B',$$

the well-known symmetric difference of events A and B. But it too fails, for instance, for all disjoint, non-complementary events A and B with positive probability. For then $P(AB) = 0$ and so $P(B|A) = 0$, but $P(AB \cup A'B') = P(A'B') = 1 - P(A \cup B) = 1 - P(A) - P(B)$, which is not 0 because A and B are disjoint but not complementary. This completes the proof of Theorem 2.2.1.

Thus if logic and probability are to be combined in a unified theory where $P(q|p)$ is the truth value of "if p then q", then an additional object and operation, "(q|p)", q given p, must be added to the usual Boolean operations of "p ∨ q", "p ∧ q", and " ~p". In terms of probabilistic events a new "conditional event" and operation "(B|A)" must be added to the operations of union (∪), intersection (∩), and complement (').

2.3 The Idea of a Conditional Event Algebra

The notion of a conditional event algebra, an algebra of fractions of propositions or events, bridges the gap between the logical versus probabilistic treatments of "if b then a". The divergence between these two developments for uncertain information is quite dramatic and constitutes a glaring partial inconsistency between logic (using the material conditional) versus conditional probability. The solution is to embed the original propositions or events in a larger system of ordered pairs (fractions) much like the way the system of integers is embedded in the larger system of all integer fractions. In this way there will be fractions (conditional propositions) that can be assigned the usual conditional probability, which is otherwise impossible while remaining inside a Boolean algebra of propositions or events. Operations corresponding to "and", "or" and "not" allow the usual manipulations; but the system is not wholly Boolean. The system can be extended to iterated fractions and can support various kinds of deductive relations between conditionals. This algebra faithfully represents uncertain information with changing context such as in all logical and probabilistic situations involving games of chance. Thus it facilitates natural language simplification, expert system rule simplification, situation assessment, data fusion, statistical contingency table analysis, Bayesian analysis and updating of conditional information, and artificial (or natural) intelligence.

2.3.1 Boole's Lost Division Operation for Conditioning

It may come as a surprise even to those who routinely use the so-called "Boolean" algebra of sets for representing and manipulating propositions in formal logic, or events in probability & statistics, to be told that you are using George Boole's algebra [7] minus a crucial development initiated by him but not part of the subsequent formalization of so-called "Boolean" algebra promoted by his successors. See Hailperin [20].

This missing development is the division of one event or proposition by another, or more properly, a new construct - a "propositional fraction" or "event fraction". The set of all such event fractions (or propositional fractions) includes the set of all simple events (or propositions) just like the ordinary number fractions include the integers namely as the subset of fractions whose denominator is 1.

The point of this crucial development is to expand the objects available so as to adequately represent the context (conditioning) in logic and in probability, something which is otherwise impossible[11] while remaining in a so-called "Boolean" algebra. George Boole had a division construct in his original work, but he died before he could adequately clarify and complete his algebra; those who came after could not follow his preliminary, non-rigorous ideas. So they closed off his algebra with just three operations and called it "Boolean" algebra. The whole subsequent development of the theories of logic and of probability has been greatly hindered by this conspicuous lack of an adequate algebraic construct to represent conditioning in both logic and probability.

2.3.2 Doing Without Fractions

The situation today is quite comparable to the early Greeks and Romans having no symbols for numerical fractions, just three integer operations of +, −, and *. Although the necessary relationships were

---

[11] In 1976 D. Lewis [61] showed that (a|b) could not be assigned the probability $P(a|b)$ and also be an element of the original Boolean algebra containing a and b because then $P(a|b) = P((a|b) \wedge a)) + P((a|b) \wedge \neg a)) = P((a|b) \mid a) P(a) + P((a|b) \mid \neg a) P(\neg a) = \ldots = 1\ P(a) + 0\ P(\neg a) = P(a)$, no matter what (except for trivial cases) the propositions a & b.

well expressed as early as Eudoxus of Cnidus [circa 400-355 b.c.] in his theory of "proportions", until modern times the lack of explicit algebraic symbols and operations has made application onerous. Imagine today how it would be were one to be told never to use numerical fractions or decimals, but only integers. Since all calculations with fractions can be reduced to integer operations without division, it is always possible to avoid all fractions. So what need is there for fractions?

Now continuing this "renunciation of all fractions" scenario, imagine trying to measure two speeds and comparing them. Not only is there the problem of assigning a whole number of distance units like miles and a whole number of time units (that usually won't come out evenly), but there is also the question of which pair (distance, time) is faster, and how much faster. Without the algebra of integer fractions (hopefully learned in fifth grade) the above problems become difficult or intractable and fraught with the likelihood of error.

Today, for the numerically literate it is no big problem to measure and compare two speeds only because we have the algebra of fractions (or machines to use it). Unfortunately, logicians and probability theorists are still laboring in analogous ways when it comes to propositions and events due to the same lack of that fourth development – division, in logic and in probability to represent conditioning, context, "if -", "given", contingencies, and a myriad of other examples of "context-sensitive" language involving propositions or events. They all require a *pair*, not just a single Boolean event, to be adequately represented especially when there are uncertainties.

Context must be carried along as part of all information just like modern numbers, which are fractions (or extensions thereof). It is not enough to count; one must also say what is counted! It is not enough to state a proposition or an event; one must also state the context, the sample space, the universe of all possibilities or models, the assumed premise of the proposition stated.

2.3.3 Probability and Logic

All this is rather strange since both logic and probability are founded on Boolean algebra: Probabilistic "events" A, B, C, ... form a Boolean Algebra just like logical "propositions" p, q, r, ... form a Boolean Algebra. Operations of "or", "and" and "not" ($\vee$, $\wedge$, and $'$) in logic correspond to "union", "intersection" and "complement" ($\cup$, $\cap$, $'$) in probability. A proposition may be true in one model and false in another just as an event may occur in one instance and not occur in another. However, whereas standard logic reduces conditionals to propositions with the same (universal) condition, probability has no explicit conditionals but does have a conditional probability based on a function P defined on its events A, B, ... having the properties:

1) $P(A \cup B) = P(A) + P(B)$, if $A \cap B = \Phi$
2) For any event A, $0 \leq P(A) \leq 1$
3) $P(\Omega) = 1$; the probability of the universal event Ω is 1.

In 1933 when A. N. Kolmogorov [27, 28] formulated his celebrated axioms for probability theory, there was no probabilistically acceptable algebra of objects "a given b" (conditional propositions) - fractions. So Kolmogorov simply defined the "conditional probability P(a|b) of a given b", as the ratio of the probability of "a and b" to the probability of "b", without defining any underlying conditional events. This has often led to difficult-to-solve probability equations without the aid of an explicit underlying algebra of conditional objects and without the direct help of deductive logic, because the latter routinely distorts the probability relationships when there is uncertainty.

In standard 2-valued logic, conditional statements like "if b then a" are routinely reduced to the proposition "a or not b", the so-called material conditional, which is not a conditional at all, but rather a proposition. (It has the universe as its implicit condition.) When proving a theorem of the form "if b then a", a mathematician can prove that in all cases "either a is true or b is false". Since mathe-matical proofs require that there be no exceptions (2-valued logic) this reduction of conditionals to simple propositions or events does not distort the logical relationships of truth. But as soon as the prop-ositions involved become uncertain, the above reduction can greatly distort the standard probabilistic measure of the partial truth of a conditional statement "if b then a", which is $P(a|b) = P(a \text{ and } b) / P(b)$.

In subsequent sections of this Chapter, the standard situations in probability and logic for conditionals will be described together with many examples of conditionals from natural language or from a simple game of tossing a six-sided die. The 3-valued truth status of conditionals will be described and used to exhibit ambiguous conditionals in common expressions. The so-called "quasi" operations of disjunction and conjunction of E. Adams [94, 95] together with the conditional negation operation comprise the first three operations of the algebra of propositional fractions or event fractions. A rigorous account of the development of integer fractions immediately precedes an analogous development of fractions for Boolean propositions or events. Features of the resulting algebra are described including extended probability laws, and the non-monotonic nature of uncertain conditionals. The four operations on conditionals give rise to four 3-valued truth tables for conditional propositions. Several sample problems and applications exhibit the utility of the algebra. Finally several types of independence between conditionals are exhibited demonstrating the descriptive richness of the algebra.

2.4 Conditional Statements With or Without Certainty
There is a very significant difference between certainty and almost certainty.

2.4.1 Statements and their Contrapositives
As G. Boole knew (see Hailperin [20]) some statements are logically equivalent when either is certain but not so when they merely have a probability. When there is certainty, a conditional, "If B then A", is equivalent to its contrapositive "If not A then not B". The converse "If A then B" is equivalent to its own contrapositive "If not B then not A".

For example, roll a 6-sided die once, let x denote the number of the face showing, and consider the conditional statement "If $x \leq 4$ then $x \leq 5$". The contrapositive of the statement is "If $x > 5$ then $x > 4$".

Note that "if B then A" is certain if and only if every instance of B is an instance of A. Therefore every instance of A', the complement of A, is an instance of B'. Thus "if not A then not B" is also certain, which is the contrapositive statement. Such inclusion relations constitute the deductive relationships existing between the various events or propositions. The conditional statement "if $x \leq 4$ then $x \leq 5$" is certain because $\{1, 2, 3, 4, 5\} \vee \{1, 2, 3, 4\}' = \{1, 2, 3, 4, 5, 6\}$ which is the universe, Ω, for rolling a 6-sided die.

But if the conditional is uncertain, then it doesn't follow that it is equivalent to its contrapositive. For example, consider the conditional statement "if $x \leq 5$ then $x \leq 4$". This has conditional probability P($(x \leq 4) | (x \leq 5)$) = P$\{1, 2, 3, 4\}$ / P$\{1, 2, 3, 4, 5\}$ = (4/6) / (5/6) = 4/5. But the contrapositive statement is "if $x > 4$ then $x > 5$" and it has the conditional probability $P((x > 5) | (x > 4)) = P\{6\} / P\{5, 6\} = (1/6) / (2/6) = 1/2$.

2.4.2 Three Truth States of Conditionals (A|B)

The reason for these ambiguities and discrepancies under uncertainty is that in any instance a conditional (A|B) can have any one of three (not just two) truth values or states:

> If A is true and B is true then (A|B) is *true*
> If A is false and B is true then (A|B) is *false*
> If B is false then (A|B) is *inapplicable*

(This is also the real reason that generations of logic students have felt uncomfortable when their teachers have explained how in 2-valued logic "if B then A" must have the same truth value as the "material conditional" proposition "A or not B", and so be equivalent to it. Someone usually objects to this reduction but is beat down by the answer that "if B then A" is false when B is true and A is false. Otherwise it is not false. So it must be true. The student knows he's been "had", but can't exactly figure out how. The reason, whose importance Bruno de Finetti [59] early recognized (1952), is that there are *three* values not just two for conditionals! The third value is "inapplicable".)

Note that for conditionals,

> "not false" is "true or inapplicable"
> "not true" is "false or inapplicable"
> "not inapplicable" is "true or false"

A conditional (A|B) is "not false" on the instances of $((A \wedge B) \vee B')$, which is the material conditional proposition $A \vee B'$. $P(AB)$ is the probability of the truth of (A|B); $P(A \vee B')$ is the probability of the non-falsity of (A|B), which goes up as the applicability of B decreases. $P(A|B)$ is the conditional probability (given B) of the truth of (A|B);

2.4.3 Cognitive Ambiguity in Uncertain Conditionals

Many common natural language conditional statements are ambiguous. This is a largely unrecognized problem. Again, this ambiguity in conditionals arises from the fact that (uncertain) conditional statements must have three, not just two, possible truth-values.

Saying "if B then A" is true is different from saying "if B then A" is not false. This ambiguity is multiplied when conditionals are combined with other conditionals or used to deduce or infer other conditionals.

Since all information is inherently conditional (has assumptions) there is often cognitive confusion when there is uncertainty and changing context. If our ubiquitous natural language conditional statements are ambiguous in meaning, then how can we hope to teach a machine to faithfully process uncertain conditional statements?

2.4.4 The Search for Coherency Between Conditional Logic and Probability

After G. Boole [7] there have been a number of serious attempts to breach the gap, but the subtlety of the algebraic problem (the habituation of "fractionless thinking" about propositions and events) seems to completely obscure the path. So often it is only afterward that we can see how hard it used to be. The Romans long used "Roman numerals" when better number systems were available. There are also very many abstract relationships in Boolean algebra that can distract the unwary researcher from the goal of finding a general common context for (conditional) logic and conditional probability.

In spite of these subtle difficulties, S. Mazurkiewicz [33-35], B. De Finetti [90], G. Schay [60], T. Hailperin [20], I. R. Goodman et al [74-75, 79], H. T. Nguyen [80] & G. S. Rogers [78] and E. A. Walker [82]), and D. Dubois & Prade, H. [83-4, 96, 97] have made valuable contributions to the field.

2.4.5 Two "Quasi-" Operations for Combining Conditionals

While there have been a number of attempts, one particular algebra seems to be most natural and immediately useful for manipulating uncertain conditional propositions and events. Just as there is a preferred or default definition for numerical fractions and preferred operations on those fractions for purposes of representing and manipulating numbers, and for expressing other operations, so too is there a preferred algebra of conditionals. It was really E. Adams [94-95], who in 1965-66 first offered completely adequate conjunction and disjunction *operations* for conditionals in one context.

Adams considers these "and" and "or" operations on conditionals to be in some sense "quasi" because they are logically and probabilistically non-monotonic. (But that is how conditional language works! See the die-tossing example of subsections 2.6.3–4.) Adams does not offer an abstract algebraic development of conditionals. But he does successfully apply his operations even to complicated two-stage, possible worlds, probability models. But his operations have a much wider application.

By 1985, after about 15 years of intermittent research, the author [9, 63] had independently rediscovered Adams' three operations (including conditional negation) and had added a fourth operation, and a new construct of division to represent conditionals, thereby extending Boole's algebra to fractions that can be combined by four operations and assigned a conditional probability. Subsequent work [64–68] in the early 1990's included a development of deduction for conditionals. Ironically, this system of conditional events is not altogether "Boolean", but it has many sub-Boolean algebras. More importantly it is rich enough in objects to have a fraction (a|b) that can be assigned the conditional probability P(a|b) and that can also be used for deduction purposes. (Boolean algebra is a misnomer anyway, because Boole was working on a division operation when he died.)

2.4.6 Conditional Conditionals

With respect to the fourth operation of conditioning and the new construct of fractions the linguistic (semantic) ambiguity due to the unrecognized 3-valued ambiguity is multiplied: For instance, consider the common iterated conditional form "If A then, if B then C". Some plausible meanings of this iterated conditional depending on the context are:

If A then, "if B then C" is not false	- ((C ∨ B') | A)
If A then, "if B then C" is true	- (C ∧ B | A)
If A and B, then C	- (C | A ∧ B)

So in the familiar probability experiment of rolling a single six-sided die once, with the die's faces as usual numbered from $x = 1$ through 6, the statement

If x is even then, if $x < 5$ then $x = 2$

plausibly corresponds in various circumstances to any one of the conditionals

$(x \in \{2,5,6\} \mid x \text{ is even})$, which has conditional probability 2/3
$(x \in \{2\} \mid x \text{ is even})$, which has conditional probability 1/3
$(x \in \{2\} \mid x \in \{2,4\})$, which has conditional probability 1/2

depending on how the statement is interpreted:

If x is even then, "if x < 5 then x = 2" is not false,
If x is even then, "if x < 5 then x = 2" is true, or
If x is even then, if also x < 5 then x = 2

The way to avoid such embarrassing anomalies is not to claim that compound and iterated conditionals are simply not linguistically or semantically well-defined, and anyway unnecessary, but rather to better specify the 3-valued truth status of conditionals in our statements and also define conditional conditionals precisely. We can take the third interpretation as default and require the further specification of truth-values if either of the first two interpretations is desired.

We are all frequently stringing conditionals together and often have no trouble assigning meaning thereto. For example, consider the following example of combining two conditionals with "or" or "and": The situation is that my friend visits me every Sunday, and passes a store on the way. "If he remembers to stop at the store, then if they have cola, he buys me a six-pack of cola, and if they have beer, he buys me a six-pack of beer." This is not a far-fetched example of an iterated and compound conditional of the form

If A, then {if B then C, and if D then E}

There are uncertainties about my friend's remembering to stop, about the store's having Coke or beer, and about his finding them in the store to buy (and having enough money to buy them). So there are uncertainties all over the place that make indiscriminate use of the material conditional reduction quite inappropriate. On the other hand, probability theory has no standard way of handling this construction either.

One of the most dramatic examples of the unrecognized use of compound conditioning was the first military strategy of our nation.

As the Colonialists waited for the British to attack, the signal was "One if by land and two if by sea". This is the conjunction of two conditionals with uncertainty! Besides the uncertainty of land or sea, there was uncertainty that the signal (one or two lamps in a window) would be given, seen and that it would be correct[12].

2.5 Development of Fractions for Integer Arithmetic

The development of fractions for propositions and events follows the analogous standard development of fractions for integer arithmetic:

Let K be the set of all integers, $\{0, -1, +1, -2, +2...\}$, under the usual operations of addition, multiplication and negation. Let R = {(a, b): a, b in K}, be the set of all ordered pairs of integers. R is called the set of rational numbers or fractions "a divided by b" of K.

As a notational convention, (a, b) is denoted (a/b) and R can be denoted (K/K).

Equivalence Relation on (K/K). Two fractions (a/b) and (c/d) are equivalent if and only if

$$\{b = 0, d = 0\} \quad \text{or} \quad \{b \neq 0, d \neq 0, \text{and } ad = bc\}$$

where "bc", denotes "integer b multiplied by integer c". That is,

$$(a/b) = (c/d) \text{ if and only if } (b = d = 0) \text{ or } (ad = bc, b \neq 0 \neq d)$$

Note: Fractions with zero denominator form their own equivalence class. They are all equivalent to each other and said to be "undefined".

The operations on fractions are well known:

$$(a/b) + (c/d) = (ad + bc \,/\, bd)$$
$$(a/b)(c/d) = (ac \,/\, bd)$$
$$-(a/b) = (-a/b)$$
$$(a/b) / (c/d) = (a / b(c/d)) = (ad / bc), \text{ \& undefined if } d=0.$$

Note: (K/K) includes the system of integers K as the fractions (a/1) for all a in K.

---

[12] This example was noticed by Dr. Alan Gordon.

2.6 Development of Fractions for Events and Propositions

Now proceeding in an analogous fashion, let B be an initial Boolean algebra of propositions or events. For example, the following are standard examples:

>B = Boolean algebra generated by any finite or infinite set of propositions
>B = Boolean algebra of subsets of a probability sample space Ω.
>B = All 64 subsets of the six-element sample space $\{1, 2, 3, 4, 5, 6\}$

Let $(B \mid B) = \{(a|b): a, b \text{ in } B\}$, which is called the set of conditionals "a given b" of B.

Next define an equivalence relation on $(B \mid B)$: Two conditionals $(a|b)$ and $(c|d)$ are equivalent if and only if

- Their conditions, b and d, are equivalent propositions or events, and
- Their conclusions, a and c, are equivalent when their common condition is true.

That is,
$$(a|b) = (c|d) \quad \text{if and only if} \quad b = d \text{ and } ab = cd$$

where juxtaposition, "ab", denotes "a ∧ b".

Note that $(a|b) = (ab \mid b)$, and the latter is said to be in reduced form. It is also easy to see that equivalent conditional propositions are assigned the same conditional probability.

2.6.1 Operations on Conditional Events

Based on semantic and other considerations (See Chapter 3 for justifications) the following operations are defined, together with some associated algebraic simplifications and semantics:

Conditional Negation: $(A|B)' = (A' \mid B)$,

The right hand side is the negation of A given B, which has (conditional) probability $1 - P(A|B)$

Disjunction: $(A|B) \vee (C|D) = (AB \vee CD) | (B \vee D)$

The right hand side means, "If either conditional is applicable then at least one is true."

Conjunction: $(A|B) \wedge (C|D) = [AB(C \vee D') \vee (A \vee B')CD] | (B \vee D)$

The right hand side means, "If either conditional is applicable then at least one is true while the other is not false".

$$(A|B) \wedge (C|D) = (ABD' \vee ABCD \vee B'CD | B \vee D)$$

That is, "If either conditional is applicable then one is true while the other is inapplicable, or both are true."

Iterated Conditioning:

$$(A|B) | (C|D) = (A | (B) (C|D)) = (A | (B (C \vee D')))$$

The right hand side means, "If B is true and (C|D) is *not false,* then A is true."

The resulting algebra $(\mathcal{B}|\mathcal{B})$ of conditional propositions or events includes the original Boolean algebra $\mathcal{B}$ as the subset of conditionals $(\mathcal{B} | \Omega)$, in which Ω is the universal event. In logical notation these are the conditional propositions $(\pounds | 1)$ whose condition is certain.

2.6.2 Venn Diagram of the Disjunction (∪) of two conditionals

The disjunction and conjunction of conditional propositions or events can be depicted with a modified Venn [57] diagram:

$$(A|B) \cup (C|D) = (AB \cup CD) | (B \cup D)$$

Two Overlapping Conditions	Union of Two Conditionals
B & D	(A|B) ∪ (C|D)

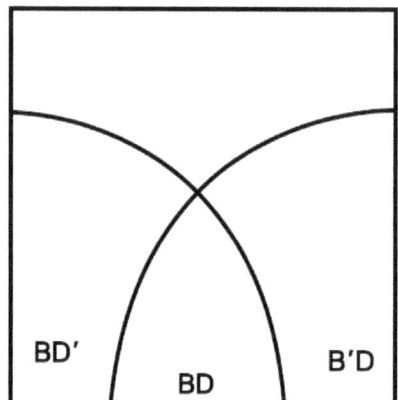 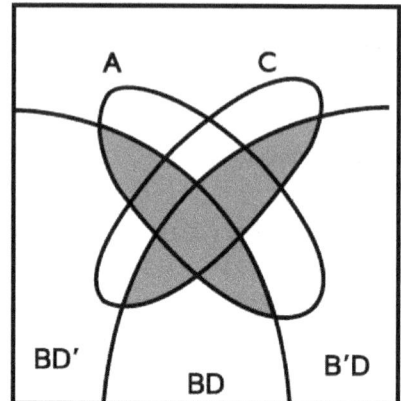

Figure 2.1 Venn Diagram of the Disjunction (∪) of Two Conditionals

The union of the 7 shaded regions corresponds to the subset AB ∪ CD, which is a subset of the union of the two conditions, B and D. (The two lowest two colored regions are those that must be removed when two conditionals are intersected in contrast to taking their union. These two subsets comprise the instances of both B and D being true but only one of A or C being true.)

2.6.3 Combining Conditional Events with "OR" (∨)

Consider the familiar probability experiment of rolling a single six-sided die once. (The die's faces are numbered from x = 1 through 6.) What is the conditional event and the conditional probability that

"If x is odd then x ∈ {1, 3, 5} OR if x is even then x = 2" ?

Were this statement translated in the usual logical way it would become the unconditional event

"[x ∈ {1, 3, 5} or not odd] OR [x = 2 or not even]"

This latter event has probability 1 since any number rolled satisfies it. But the original statement is false for a roll of 4 or 6.

On the other hand, by this algebra of conditionals, the initial statement can be written

$$(x \in \{1,3,5\} \mid x \text{ is odd}) \vee (x = 2 \mid x \text{ is even})$$
$$= (\text{odd} \mid \text{odd}) \vee (\{2\} \mid \text{even})$$
$$= (\text{odd} \vee \{2\} \mid (\text{odd or even})$$
$$= (\{1,2,3,5\} \mid \Omega) = \{1,2,3,5\}.$$

And the (conditional) probability is correctly calculated as 4/6.

2.6.4 Non-Monotonicity of Probability on Conditionals

The above example well illustrates how probability is non-monotonic when operating on conditionals. In the above situation of rolling a single die once, were I to make just the first part of the statement, namely "if x is odd then x is 1, 3, or 5" then nobody would argue with me because the statement would be perfectly true and have conditional probability 1. But, if I don't stop there, and instead continue with "or if x is even then x = 2", now people will argue with me because rolls of 4 and 6 will be applicable, and not satisfy my statement.

It seems rather shortsighted of people (who would perhaps like to remain fractionless) to argue that because constructions such as "if b then a, or if d then c" are supposedly semantically ambiguous, they should remain so. The fact is that everyday life abounds with examples of just these kinds of compound conditionals. For instance, even so common a rule as "heads you win, and tails you lose" is a compound conditional.

2.6.5 Default Interpretation of Iterated Conditionals

A good way to view conditional conditionals is by interpreting "c given d", where c & d may be conditionals, as "c given that d is not false" as contrasted from "c given that d is true". In the Boolean case these two are equivalent. But in the realm of conditionals, they are no longer equivalent. Thus "a given that (c|d) is true" is not the same as "a given that (c|d) is not false" due to the third truth state of "inapplicable". "a | (c|d) is true" is equivalent to "a given (c ∧ d) is true", namely (a | c ∧ d), which is also equivalent to "((a | c) | d)", but "a given that (c|d) is not false" is (a | c ∨ d′).

2.7 Truth Value Representation

Since the propositions or events of this theory are ordinary 2-valued (crisp versus fuzzy) propositions, this algebra of 3-valued conditionals is completely defined by the following truth tables, which can be computed directly from the operations, where the third value of "inapplicable" is represented as I:

	AND T F I	OR T F I	GIVEN T F I	NOT
T	T F T	T T T	T I T	F
F	F F F	T F F	F I F	T
I	T F I	T F I	I I I	I

Table 2.1 Three-Valued Truth Tables

Note here again that "not true" means "false or inapplicable", that "not false" means "true or inapplicable", and that "true" and "false" are no longer complete opposites.

2.8 Partial Solution to an Expert System Circularity Problem

Back when rule-based expert systems were becoming popular, a recurrent problem was a set of rules that, when chained, could cause a circular computer search. For instance[13], denote the following propositions as indicated:

T = proposition that patient has a temperature > 100° F
F = proposition that patient has a fever
S = proposition that patient has flat pink spots
M = proposition that patient has measles.

Rule 1: If T   then F
Rule 2: If (F and S)   then M
Rule 3: If M   then T

---

[13] On p. 72 of "Knowledge Base Verification", Tin A. Nguyen, Walton A Perkins, Thomas J Laffey, and Deanne Pecora, *AI Magazine*, Summer 1987, 69-75.

Clearly, since Rule 3 refers back to an event T in Rule 1, if these rules are chained, then the computer program may never end! Using conditional operations (and skipping some Boolean algebra simplification steps) the conjunction of the 3 rules can be simplified:

$$(F\mid T) \wedge (M\mid FS) \wedge (T\mid M)$$
$$= (F\mid T) \wedge (T\mid M) \wedge (M\mid FS)$$
$$= ((FT) \mid (T \vee M)) \wedge (M\mid FS)$$
$$= (FTM(FS)' \vee (T'M')MFS \vee FTMS) \mid (T \vee M \vee FS)$$
$$= (FT(S' \vee M) \mid (T \vee M \vee FS)).$$

So the conjunction of the three rules is logically and probabilistically equivalent to a single rule: If the patient has a temperature over 100 or measles or both a fever and flat pink spots, then the patient has a fever, a temperature over 100, and either measles or no flat pink spots. By thus combining the rules, there may be no need to chain and no danger of an infinite loop.

However, it will be shown in Chapter 6 that when conditional propositions are not certain, the conjunction of two conditional propositions does not necessarily *imply*[14] either of the component conditionals! This is non-monotonicity. In general it is therefore necessary to retain the individual component conditionals and their conjunctions in determining the logical and probabilistic consequences of a set of conditionals with given conditional probabilities.

2.9 A Problem of Iterated Conditioning

Ernest Adams has posed an interesting problem in Bayesian updating: An object of unknown color may be red (r), yellow (y) or blue (b) with equal probability. What is the new probability of blue upon learning that 'if the object is not red it is blue'? Thus, what is $P(b \mid (b\mid r'))$ and more importantly, how should this probability be calculated?

The difficulty is that the new information is itself conditional. So it is not an event upon which to condition in the Bayesian sense. But using this algebra of conditionals, upon learning that "if the object is not red then it is blue", the new (conditional) event for "blue (b)" is

---

[14] For conditionals there are a variety of plausible deductive relations competing for the word "imply".

$$(b \mid (b \mid r')) = (b \mid (b \vee r)),$$

and the new probability for b is $P(b \mid (b \vee r)) = P(b) / P(b \vee r) = (1/3) / (2/3) = 1/2$.

This example well illustrates how the material conditional reduction, $(b \vee r)$, of the conditional $(b \mid r')$ is consistent with the new algebra when that conditional is itself a condition of an iterated conditional. Standard Boolean logic, including the material conditional simplification of true conditionals, can proceed as usual *within the hypothesized premises of a proposition* while the conjectured conclusions remain probabilistic.

Thus logic (that which is true in all or no cases) is not distorted and confounded by probability (that which can be true in some cases.)

2.10 Probabilities of Disjunctions and Conjunctions of Conditionals

The following are extensions of the usual laws for expressing the probability of the union and intersection of events:

$$P((A|B) \vee (C|D)) = P(B|B \vee D)P(A|B) + P(D \mid B \vee D)P(C|D) - P(ABCD \mid B \vee D)$$

$$P((A|B) \wedge (C|D)) = P(B|B \vee D)P(AD' \mid B) + P(D|B \vee D)P(CB' \mid D) + P(ABCD \mid B \vee D)$$

Note that the last term $P(ABCD \mid B \vee D)$ can naturally be expressed as the product $P(BD \mid B \vee D) P(AC \mid BD)$, the probability that both conditions are true given either is, times the probability that both conclusions are true given both conditions are.

2.11 Overlapping surveillance regions

Let "a" be detection by one sensor having surveillance region "b". Let "c" be the event of detection by a second sensor with region "d". Then detection by either radar is the union of conditionals, (a|b) ∪ (c|d), that if the object is in region b it will be detected by sensor 1, or if the object is in region d then it will be detected by sensor 2.

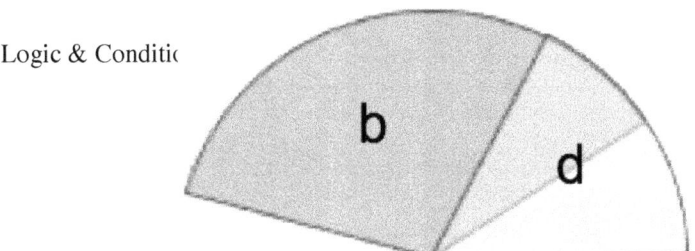

Figure 2.2 Surveillance Regions

For Simplicity, assume the object moves randomly over regions b and d, and that the area of region b is 9 square units and the area of region d is 4 units, with 2 units of area overlap.

P((a|b) ∪ (c|d)) = P(a|b) P(b | b ∪ d) + P(c|d) P(d | b ∪ d)
  - P(ac | bd) P(bd | b ∪ d)
  = P(a|b) (9/11) + P(c|d) (4/11) - P(ac | bd) (2/11)

If detection by sensor 1 is independent of detection by sensor 2 given the object is in the overlapping region bd, that is, if a is conditionally independent of c given bd, then P(ac | bd) = P(a | bd) P(c | bd). Additional conditional independence of a and d given b, and of c and b given d, yields

P((a|b) ∪ (c|d)) = P(a|b)(9/11) + P(c|d)(4/11) - P(a | bd)P(c| bd)(2/11)
  = P(a|b)(9/11) + P(c|d)(4/11) - P(a|b) P(c|d)(2/11).

Conditional independence is a popular way to simplify Bayesian nets. See J. Pearl [81].

2.12 Simpson's Reverse Paradox

One of the most revealing examples of the general lack of adequate understanding of fractions (conditionals) as applied to propositions and events occurs in a myriad of statistical contexts - recently called "Simpson's reverse paradox". Consider the following data depicted below concerning a sample space of male and female patients suffering from a certain disease who receive or do not receive a medicine and get better or not.

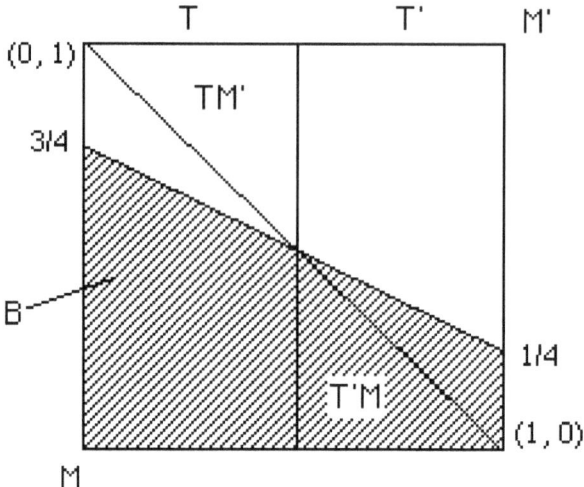

Figure 2.3 Simpson's Paradox

$$P(B \mid T) > P(B \mid T')$$
$$P(B \mid TM) < P(B \mid T'M)$$
$$P(B \mid TM') < P(B \mid T'M')$$

In Figure 2.3 the probability (area) that patients with a certain disease get better (B) given treatment (T) is higher than without treatment (T'), but not for men (M) nor for women (M'). This is Simpson's (Reverse) Paradox!

That such an apparent anomaly seems "impossible" to so many people in this and many other contexts demonstrates a certain naïveté that we all have about how conditionals operate. But suppose the same "paradox" is rephrased, say, in terms of the batting averages of two players A and B in two years 1 and 2. Due to the variable number of at-bats of each player in each year, it is quite easy to imagine player A having a much better batting average (hits per number of at-bats) than player B in both years, but not at all when both years are combined. But these are essentially the same situations except for the terminology: The men versus the women patients correspond to the two baseball years; the medicine or not corresponds to player A versus player B; the survival rates of the men and women correspond to the batting averages in the two years. In an unfamiliar situation, our intuition for conditional thinking seems to disappear almost like the number "two" used to when people had a different word for each kind of pair, almost like fractions disappear in the mind of a fifth grader before he finally "gets it". It is that subtle.

2.12.1 Conditional Event Analysis of Simpson's Paradox

Use of the algebra of conditionals allows a natural development and promises to lead to a new statistical test of the significance level of the occurrence of Simpson's paradox in its various visitations. For instance, for the medical example

$$(B|T) = (B \mid TM) \cup (B \mid TM')$$
$$= (B \mid TM)(M \mid T) \cup (B \mid TM')(M' \mid T)$$

$$(B|T') = (B \mid T'M) \cup (B \mid T'M')$$
$$= (B \mid T'M)(M \mid T') \cup (B \mid T'M')(M' \mid T')$$

$$P(B|T) = P(B \mid TM)P(M \mid T) + P(B \mid TM')P(M' \mid T)$$
$$P(B|T') = P(B \mid T'M)P(M \mid T') + P(B \mid T'M')P(M' \mid T')$$

A necessary condition for $P(B|T) > P(B|T')$ to hold is that at least one of the factors being weighted in the equation for $P(B|T)$ must be greater than at least one of the factors being weighted in the equation for $P(B|T')$. If these reverse inequalities hold then this condition is also sufficient for Simpson's "Reverse Paradox" to occur since unit weights easily exist for which $P(B|T) > P(B|T')$.

The same argument applies for a partition $\{M_i: i = 1, 2... n\}$ of the sample space of all patients with the disease.

$$P(B|T) = P(B \mid TM_1)P(M_1 \mid T) + P(B \mid TM_2)P(M_2 \mid T) + ...$$
$$+ P(B \mid TM_n)P(M_n \mid T)$$
$$P(B|T') = P(B \mid T'M_1)P(M_1 \mid T') + P(B \mid T'M_2)P(M_2 \mid T') + ...$$
$$+ P(B \mid T'M_n)P(M_n \mid T')$$

Necessary & Sufficient Condition for the *possibility* of Simpson's paradox is:

$$\text{Max } \{P(B \mid T\ M_i)\} > \text{Min } \{P(B \mid T'\ M_i)\}.$$

2.12.2 A Statistical Question
The probability of Simpson's Paradox in various weighting circumstances can now be considered on the way to a statistical test. For instance in the medical context, using any fixed set of four weights for the ratios of the number of men (women) to the total number of persons treated (not treated), and given that more people with treatment survive than without, what is the probability that the "reverse" will occur? If this probability, over all partitions with these weights, is greater than 5%, then data having those weights cannot be used to conclude with 95% confidence that treatment would be beneficial to anyone. With those weights, there might be too great a chance ($> 5\%$) that treatment would worsen the condition of both men and women or worsen the condition of all subclasses of some other partition of the population with those weights. Can Simpson's Paradox be statistically quantified in this way using the algebra of conditionals?

2.13 Cooperative Targeting Hypothesis Problem
To show that the algebra offered herein is not just for toy problems with dice, consider a complicated problem in military situation assessment: While naval ships at sea have a lot of time to react and defend against an attacking aircraft within radar line of sight, it is possible for an enemy aircraft to launch a low missile from over the horizon aimed at a surface ship. Such a missile might not be observed until it was too late. But this requires targeting information to be passed to the attack aircraft from some other source like a cooperating (C) high altitude surveillance aircraft.

Information on this latter surveillance aircraft may develop from its simple detection (S) through the designations of "non-friendly" (N), "hostile"(H), "target (that's us!) acquired" (A), and "passing data" (g) phases. In the meantime, intelligence information on the attack aircraft may also gather from simply knowing that such an aircraft is in the area (a) to knowing that it has line of sight communications (w) with its cooperating surveillance aircraft so that data passing might occurred.

Thus this developing situation offers several precursor events for a surface ship to observe in assessing the likelihood of such an attack at any moment and so alert the ship's captain of the imminent danger.

A typical scenario would involve information from friendly AAW intelligence (Anti-Air Warfare) and Elint radar sources, own ship observations and whatever deductions and inferences could be made in time to aid the human decision maker.

Various logical relationships between the events such as "every hostile (H) is an unfriendly (N)" are expressed with standard Boolean relations (H ≤ N).

Here is a real example problem that has been solved using this new algebra of fractions. See Cooperative Targeting Hypothesis (Figure 2.4) together with the following events:

   C - Cooperative targeting of our surface ship by hostile surveillance and attack aircraft
   S - Surveillance radar operating in area
   N - Non-friendly, high altitude, inbound aircraft detected
   H - Hostile surveillance aircraft in area identified
   A - Targeting data on our ship has been acquired
   g - Passing target data has occurred
   u - AAW report of unknown air track
   r - Elint radar intercept (of surveillance radar operating in area)
   t - Elint radar hit of surveillance radar in tracking mode
   i - Elint communications intercept of target data passing
   j - AAW Intelligence of attack capable aircraft in area
   w - Attack-capable aircraft in line of sight of surveillance aircraft
   a - Attack capable aircraft in area

The following Boolean deductions (≤) hold:

$$C \leq w, \quad C \leq aA, \quad C \leq aH$$
$$w \leq ag \leq a, \quad ag \leq g$$
$$g \leq A \leq H \leq N, \quad H \leq S$$

Initial Analysis:
$$C = (C \mid aH) \vee (C \mid (aH)')$$
where
$$(C \mid aH) = (C \mid aHA' \vee aHA)$$
$$= (C \mid aHA' \vee aA)$$
$$= (C \mid aHA' \vee aAg' \vee aAg)$$
$$= (C \mid aHA' \vee aAg' \vee ag)$$
$$= (C \mid aHA' \vee aAg' \vee agw' \vee agw)$$
$$= (C \mid aHA' \vee aAg' \vee agw' \vee w)$$

$$= (C \mid aHA') \vee (C \mid aAg' \vee agw') \vee (C \mid w)$$

Thus

$$C = (C \mid aHA') \vee (C \mid aAg' \vee agw') \vee (C \mid w) \vee (C \mid (aH)')$$

So

$$P(C) = P(C \mid aHA') P(aHA') + P(C \mid aAg' \vee agw') P(aAg' \vee agw')$$
$$+ P(C \mid w) P(w) + P(C \mid (aH)') P((aH)')$$

Here, as depicted in Figure 2.4, $P(C \mid aHA')$ is "low", $P(C \mid aAg' \vee agw')$ is "medium", and $P(C \mid w)$ is "high".

Further partitioning of these events allows a full expression of the probability of a cooperative targeting event given various precursor events. The ability to manipulate events and conditional events without distorting the associated conditional probabilities greatly facilitates the solution of such problems.

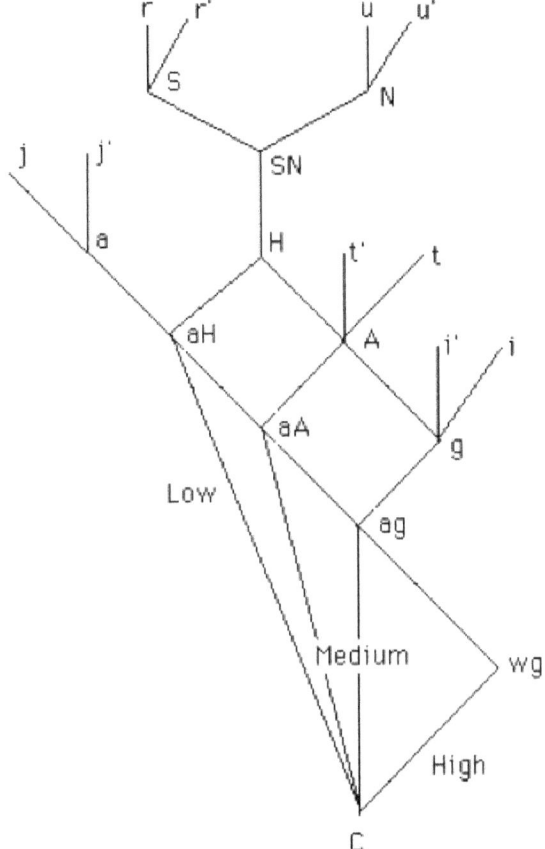

Figure 2.4 Cooperative Targeting Hypothesis

2.14 Independence and Conditionals

For Boolean propositions or events a and b, *independence* means that $P(ab) = P(a) P(b)$, which is equivalent to $P(a \mid b) = P(a)$ and to $P(b \mid a) = P(b)$. The occurrence of b does not affect the probability of a, and vice versa.

However, as with deduction between conditionals, *independence* between conditionals can naturally take on more than one plausible meaning (due to there being two components to a conditional and three truth values). Among these types of independence there are:

2.14.1 Definition (Conditional Independence)

A proposition a is independent of a proposition b given a third proposition c if and only if $P(ab \mid c) = P(a \mid c)P(b \mid c)$. This is easily equivalent to $P(b \mid ac) = P(b \mid c)$ and also to $P(a \mid bc) = P(a \mid c)$.

2.14.2 Truth Independence
A conditional event (a|b) is *truth independent* of a conditional event (c|d) provided the event (ab) is independent of event (cd). That is, $P(abcd) = P(ab)P(cd)$.

2.14.3 Applicability Independence
A conditional event (a|b) is *applicability independent* of a conditional event (c|d) provided the event b is independent of event d. That is, $P(bd) = P(b)P(d)$.

2.14.4 Falsity Independence
A conditional event (a|b) is *falsity independent* of a conditional event (c|d) provided the event (a'b) is independent of event (c'd). That is, $P(a'bc'd) = P(a'b)P(c'd)$.

2.15 Summary
Commonly recognized sources of ambiguity in language include:

- Uncertainties of occurrence such as "an even number is rolled"
- *Fuzzy* meanings such as "a *large* number is rolled"
- Confusion of the converse with the contrapositive and other logical errors.

Little recognized sources of large amounts of cognitive ambiguity: (Often the credit for these ambiguities is given to the above sources.)

- Inadequate representation of conditional statements
- Ambiguous interpretation of conditional statements
- Inadequate processing of conditional statements

The lack of conditionals in logic and probability is like being without integer fractions when doing numerical calculations --- possible, since all calculations can be reduced to integer operations, but very awkward and fraught with the likelihood of miscalculation.

Logic and probability need the common context of event fractions, 3-valued propositional fractions.

Chapter 3

Logic and Probability Synthesized in Algebraic Logic and Model Theory

The truth definition is the bridge connecting the formal language with its interpretations by means of models. If the truth-value 'true' goes with the sentence φ and model m, we say that φ is true in m and also that m is a model of φ.
C. Chang & H. Keisler [14], *Model Theory*, 1973.

3.1 Truth Values Versus Probabilities

The truth values of the two-valued logic have sometimes been introduced in terms of a homomorphism h from the Boolean algebra set of propositions L into the two-element Boolean algebra {0,1}. That is, propositions are each assigned truth-value 1 (true) or 0 (not true, false) according to the following 3 rules:

$$h(p \wedge q) = h(p) \wedge h(q) \qquad (3.1.a)$$

That is, (p ∧ q) is always true if and only if both p is always true and q is always true.

$$h(p \vee q) = h(p) \vee h(q) \qquad (3.1.b)$$

That is, (p ∨ q) is always true if and only if either p is always true or q is always true, or both. Thirdly,

$$\underline{h}(\sim p) = \sim(h(p)) \qquad (3.1.c)$$

That is, the truth-value of ~p is the negation of the truth-value of p. ~p is also written p′ or p'. The following 3 tables display the truth-values assigned by negation ('), conjunction (∧), and disjunction (∨) for propositions that are either True (1) or False (0).

One-place Truth Table for Negation ('):

p	p'
1	0
0	1

Two-place Truth Table for Conjunction (∧):

∧	1	0
1	1	0
0	0	0

Two-place Truth Table for Disjunction (∨):

∨	1	0
1	1	1
0	1	0

In the two-valued logic, the kernel T of this homomorphism, the inverse image of 1, is a maximal (sum) ideal of propositions, an ultrafilter of £. [It easily follows from Equation (3.1.c) that for every proposition p, either p ∈ T or ~p ∈ T.] However, if the kernel of the homomorphism h is not maximal, then there are propositions p ∈ L for which neither p nor its negation ~p is in T. This is just the right algebraic context for probability logic.

3.2 The Extension of a Proposition

If a logic £ has a truth homomorphism h and the kernel T is not maximal, then the homomorphic image of L cannot be the two-element Boolean algebra {0,1}. We have in mind, instead, the real interval [0,1], which includes the images of all probability measures. Furthermore the homomorphic properties of Equations 3.1.a and 3.1.b are not appropriate for the assignment of probabilistic truth-values. The former would have all events (stochastically) independent. The latter would have all events disjoint.

In algebraic logic the homomorphism h becomes (see Theorems 3.8.3 and 3.8.4) the extension (or model) function m:

Definition 3.2.1. If I is a set of propositions of a Boolean logic £, and Ω is a well-defined universe of models that satisfy the axioms of £, then the extension of I, m(I), is just the set of all models satisfying the propositions of I. That is,

$$\text{m}(I) = \{\omega \in \Omega : \text{all propositions of } I \text{ are true in to } \omega\} \quad (3.2.\text{a})$$

For a proposition p this becomes

$$\text{m}(p) = \{\text{models } \omega : p \text{ is true in } \omega\} \quad (3.2.\text{b})$$

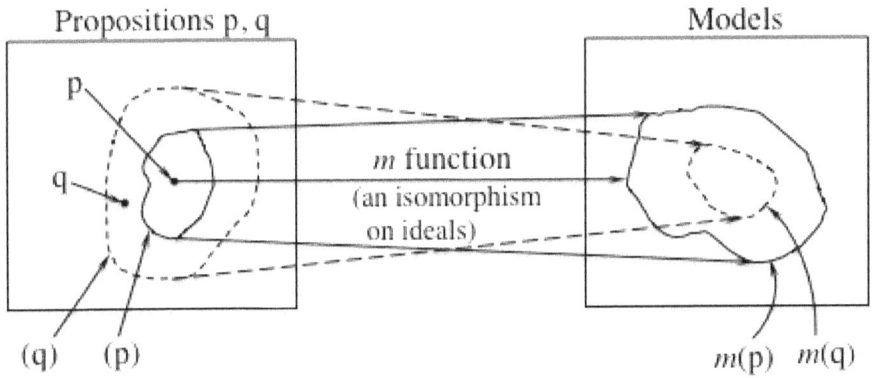

Figure 3.1 Propositions Filtering Models via the Extension Function m

The extension m(p) is just the set of models ω in which p is true. Thus the truth of a proposition p can be defined in terms of the truth of p in examples that satisfy the premise (implicit conditions) of p. Probabilistic truth values can then be assigned to the extensions of propositions rather than directly to the propositions by P(p) = P(m(p)). Thus m(p) must be the P-measurable subset of all models in which p is true.

Putting off further discussion of ideals, let us turn to defining such a probability function for the development of conditional probability logic (CPL)

3.3 "if - then - " in Logic Versus Probability

Although George Boole, the recognized father of Boolean algebra, incorporated fractions of propositions in his pioneering work [7], the parallel development of abstract algebra at the time had not proceeded far enough for Boole's immediate followers to make much sense of his far-seeing ideas about division of propositions in regard to conditioning. Consequently, those who pursued his research topics decided to close his system with just his logical counterparts to addition, multiplication and negation, and to call the result "Boolean" algebra. But Boole himself also had the division operation! Boole was explicitly trying to incorporate probability as well as logic into his system.

Prior to 1933, when A. N. Kolmogorov [27, 28] first published his celebrated axioms for probability theory, he had found that the standard treatment of "if - then -" in logic was inconsistent with the laws of probability. Since there was no probabilistically acceptable algebra of "if - then –" (conditional propositions), Kolmogorov simply defined a conditional probability without defining any underlying conditional propositions. Nor did he define Boolean-like operations on such conditionals. To this day there remains this remarkable breach in most treatments between conditional logic and conditional probability with respect to "if - then –" statements: In standard 2-valued logic, conditional statements like "if b then a" are routinely reduced to the statement "either a or not b", the so-called material conditional. For instance, when proving a theorem of the form "if b then a", a mathematician can prove that in all cases "either a is true or b is false". Since mathematical proofs require that there be no exceptions (2-valued logic) this reduction works out fairly well. But as soon as the propositions involved become uncertain, the above reduction can greatly distort the standard probabilistic measure of the partial truth of a conditional statement "if b then a".

Being aware of the above logic-probability breach, various authors[15] have attempted to define operations on conditional propositions that are consistent with both logic and probability. These efforts have resulted in several different algebras of conditional propositions

---

[15] G. Boole [7] (1854), S. Mazurkiewicz [33-35] (1932-), B. De Finetti [90, 59] (1936-), E. Adams [2, 94, 162, 153] (1966-), G. Schay [60] (1968), P. Calabrese [9, 63-67] (1975-), T. E. Hailperin [20] (1976-), N. Nilsson [62] (1986), I. R. Goodman [74-77, 79, 161] (1990-), H. T. Nguyen & G. S. Rogers [78] (1991-) and E. A. Walker [82] (1991-)

and also several ways to represent conditional propositions, an area of research that has recently been called conditional probability logic (CPL).

3.4 Formal Development of Conditional Probability Logic (CPL)

The development of the logical operations for CPL can be accomplished most simply in terms of partially defined, measurable indicator (characteristic) functions on a sample space of models (instances). This approach was early proposed by B. De Finetti [90, 59] and later utilized by G. Schay [60]. (Also see P. Calabrese [63], p. 234 and especially [64], pp. 684-686, and [65], pp. 75-82.) Alternatively, the development of CPL operations can be defined in terms of ordered pairs of propositions. For the latter development, see P. Calabrese [63], pp. 203-214. First the connection between probabilistic events and logical propositions will be described.

3.4.1 Events and Propositions

Let $\mathcal{P} = (\Omega, \mathcal{B}, P)$ be a probability space of individual instances Ω, events $\mathcal{B}$ (an algebra of subsets of instances), and probability measure P. Then the characteristic function of each measurable subset B, $B \in \mathcal{B}$ is a unique measurable indicator function q_B: $\Omega \to \{0,1\}$ from Ω to the 2-element Boolean algebra $\{0,1\}$ defined as follows:

$$(q_B)(\omega) = \begin{cases} 1, \text{ if } \omega \in B \\ 0, \text{ if } \omega \in B' \end{cases}$$

The function q (dropping the subscript) is a "proposition" in the sense that for each $\omega \in \Omega$, either q is true for ω, meaning $q(\omega) = 1$, or else q is false for ω, meaning $q(\omega) = 0$. L will denote the set of all propositions of $\mathcal{P}$. Conversely, each measurable indicator function q defines a unique measurable subset B, $B \in \mathcal{B}$ by

$$B = q^{-1}(1) = \{\omega \in \Omega : q(\omega) = 1\}$$

B is the measurable subset of cases (instances) for which q is true, and P(B) is the probability measure of the partial truth of q.

In this correspondence between measurable subsets (i.e., probabilistic events) and measurable indicator functions (i.e., propositions), the universe of all possible cases Ω corresponds to the unity indicator function, to those propositions that are true in all cases --- necessary & provable. The empty set Φ corresponds to the zero indicator function, to those propositions that are false in all cases --- impossible and contradictory. Two propositions p and q are equivalent if and only if they are equal as functions.

3.4.2 Definition of Equivalence

Two propositions (indicator functions) p and q are equivalent if and only if they are equal as functions. That is, $p = q$ if and only if both p and q take the value 1 (or 0) on the same subset of Ω. Thus $p = q$ if and only if $p^{-1}(1) = q^{-1}(1)$ if and only if $p^{-1}(0) = q^{-1}(0)$.

3.4.3 Boolean Operations on Propositions

The Boolean operations of union ($\cup$), intersection ($\cap$) and complement ($'$) defined on the Boolean algebra (or sigma-algebra) $\mathcal{B}$ of events of $\mathcal{P}$ naturally induce Boolean operations on the propositions of L: For arbitrary events A and B in $\mathcal{B}$, or propositions p_A, p_B in L, define

$$p_A \vee p_B = p_{(A \cup B)}$$
$$p_A \wedge p_B = p_{(A \cap B)}$$
$$\sim(p_A) = p_A{'}$$

This is equivalent to the equations

$$(p \vee q)(\omega) = p(\omega) \vee q(\omega)$$
$$(p \wedge q)(\omega) = p(\omega) \wedge q(\omega)$$
$$\sim p(\omega) = (p(\omega))'$$

where the operations of disjunction ($\vee$), conjunction ($\wedge$) and negation ($'$) on the right hand side of these equations are in the 2-element Boolean algebra $\{0,1\}$. $\mathcal{L}$ will denote the Boolean algebra of propositions L as generated by the probability space $\mathcal{P}$.

3.5 Conditional Events and Conditional Propositions

Consider now that each ordered pair, (B|A) of measurable subsets B, A in $\mathcal{B}$ with corresponding indicator functions q, p, defines a unique *domain-restricted* measurable indicator function (q|p): A → {0,1}, from A to the 2-element Boolean algebra as follows:

$$(q|p)(\omega) = \begin{cases} 1, \text{ if } \omega \in B \cap A \\ 0, \text{ if } \omega \in B' \cap A \\ U, \text{ if } \omega \in A' \end{cases}$$

where U means "undefined". This can be expressed in terms of the unconditioned propositions p and q by

$$(q|p)(\omega) = \begin{cases} q(\omega) \text{ if } p(\omega) = 1 \\ U \text{ if } p(\omega) = 0 \end{cases}$$

(q|p) is a "conditional proposition" in the sense that if p is true for ω then either (q|p) is true for ω or (q|p) is false for ω. But we say that (q|p) "does not apply" (i.e., is undefined or inapplicable) for those ω for which p is false. Thus (q|p) has three truth states. (q|p) is simply q, restricted to $p^{-1}(1)$, the subset on which p is true. The set of all conditional propositions of $\mathcal{P}$ will be denoted £/£.

Conversely, each domain-restricted, measurable indicator function (q|p) defines a unique conditional event (B|A), where A and B are measurable subsets determined by $A = p^{-1}(1)$ and $B = q^{-1}(1)$. A is the measurable subset on which p is true; B is the measurable subset on which q is true, and B ∩ A is the measurable subset on which both q and p are true. For non-zero P(A), P(B|A) = P(B ∩ A) / P(A) is the conditional probability of q given p, which is denoted P(q|p).

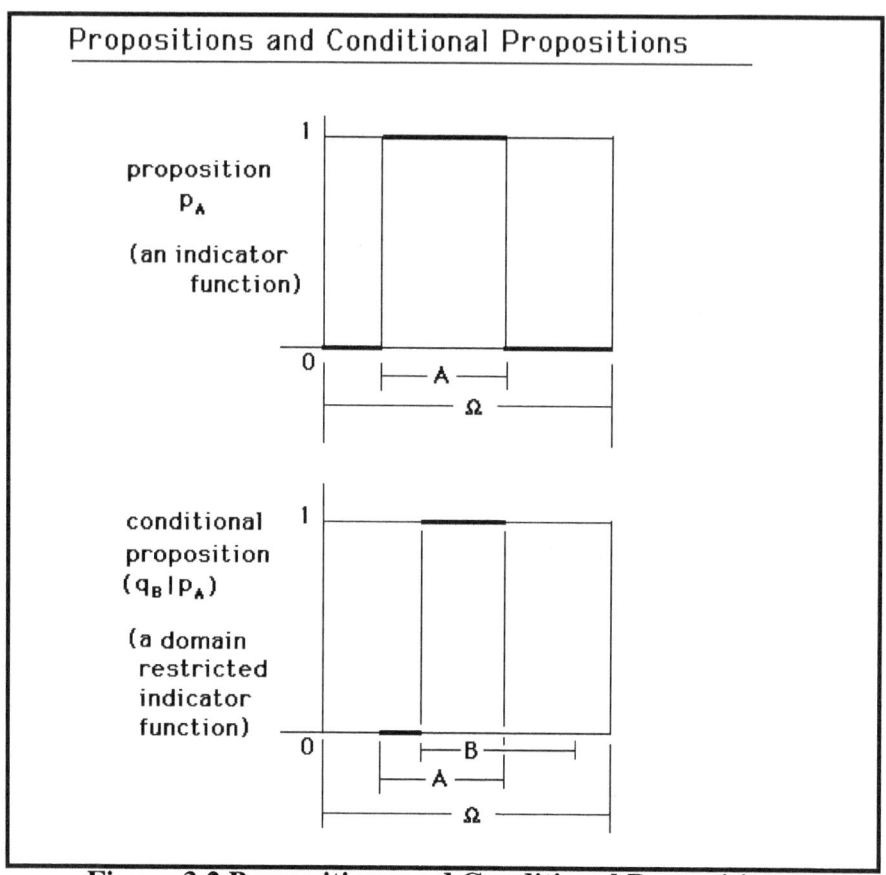

Figure 3.2 Propositions and Conditional Propositions

3.6 Definition of Equivalence for Conditional Propositions

Two conditional propositions (q|p) and (s|r) are equivalent, i.e. (q|p) = (s|r), if and only if they are equal as indicator functions, that is, if and only if they have the same domain and are equal on this common domain. Note that it easily follows that two conditionals (q|p) and (s|r) are equivalent (=) if and only if they have equivalent premises and their conclusions are equivalent in conjunction with that premise. That is,

$$(q|p) = (s|r) \text{ if and only if } (p = r) \text{ and } (qp = sr)$$

Note that if (q|p) is an arbitrary conditional then (q|p) = (qp|p). A conditional proposition (q|p) is said to be in *reduced form* if q = qp. It is also easy to see that equivalent conditional propositions are assigned the same conditional probability.

Note also that for all propositions q, (1|0) = (q|0) = (0|0). (1|0) is the "inapplicable" or "undefined" conditional and is denoted U. The Boolean propositions 1 and 0 propositions are represented by (1|1) and (0|1) respectively.

Thus, in any instance ω a conditional (a|b) can have any one of 3 truth values:

(a|b) is *true*   if (a|b) = (1|1) = 1, i.e., if a is true and b is true
(a|b) is *false*   if (a|b) = (0|1) = 0, i.e., if a is false and b is true
(a|b) is *inapplicable* if (a|b) = (1|0) = U, i.e., if b is false

Thus, (a|b) is *true* on $a \wedge b$, *false* on $a' \wedge b$, and *inapplicable* on b'.

Note also that "not true" means "false or inapplicable"; "not false" means "true or inapplicable"; "not inapplicable" means "true or false". CPL provides clear distinctions in terminology. For example, "if a then b" is "not false" on the instances of $a \vee b'$, but it is "true" on the smaller set of instances of $a \wedge b$. No such distinctions are available in purely 2-valued logic.

Having defined conditional events, the probability of the truth of a conditional event (a|b) given the truth of the premise of the conditional is defined to be the usual conditional probability P(a|b) = P(a and b)/P(b).

3.7 Operations on Conditionals
Each of the three operations defined below agrees with the corresponding Boolean operation when applied to conditionals with equivalent conditions. Therefore they extend the Boolean operations. They will first be intuitively explained.

3.7.1 Conditional Negation
The conditional negation of "a given b" is the "negation of a, given b". That is,
$$(a|b)' = (a' \mid b),$$

and the latter has probability 1 - P(a|b).

3.7.2 Disjunction
Concerning disjunction, "If b then a, or if d then c" means "If either conditional is applicable then at least one is true". That is,

$$(a|b) \vee (c|d) = (ab \vee cd) | (b \vee d)$$

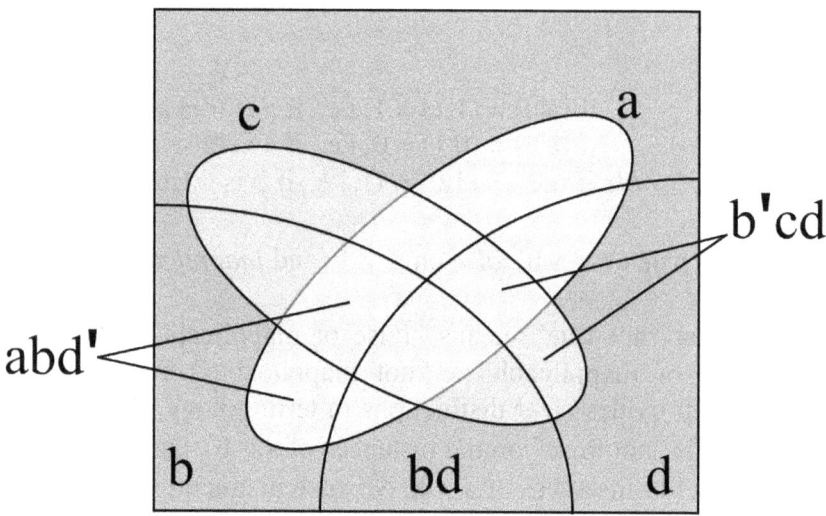

Figure 3.3 Disjunction and Conjunction of Conditional Events (a|b) and (c|d)

3.7.3 Conjunction
Concerning conjunction, "If b then a and if d then c" means "if either conditional is applicable then at least one is true while the other is not false". That is,

$$(a|b) \wedge (c|d) = [\,ab(c \vee d') \vee (a \vee b')cd\,] | (b \vee d)$$
$$= (abd' \vee abcd \vee b'cd \,|\, b \vee d),$$

which also means "if either conditional is applicable then either they are both true or else one is true while the other is inapplicable." The latter formula also follows by the standard De Morgan formulas in Boolean algebra relating conjunction and negation to disjunction and negation.

The algebra ($\mathcal{B} \mid \mathcal{B}$) of conditional events includes the original Boolean algebra $\mathcal{B}$ as those conditionals (B | Ω), where Ω is the universal event. In logical notation these are the conditionals (a |1) whose condition is certain. Analogously, these are like the integer fractions whose denominators are 1. The proposition "a" is identified with the conditional (a | 1). Fixing the condition b also yields a Boolean algebra ($\mathcal{B}$ | b).

3.7.4 Iterated Conditioning

A conditional (c|d) may itself be a condition for another proposition or conditional proposition. Due to the largely unrecognized third truth status of conditional statements, to a great degree, natural language is ambiguous about such iterated conditioning. However, here is iterated conditioning as will later be justified:

$$[(a|b) \mid (c|d)] = [a \mid b \, (c|d)] = [a \mid b(c \vee d')]$$

where the second equivalence follows since it can be shown from the first 3 operations that,
$$b \wedge (c|d) = (b|1) \wedge (c|d) = b \, (c \vee d')$$

That is, "(a|b) given (c|d)" means "a given that b and (c|d) are not false".

Conditionals are to logic as fractions are to arithmetic. As the system of integer fractions extends the system of whole numbers, so also does the system of conditionals (ordered pairs of propositions) extend the underlying Boolean algebra of propositions or probabilistic events allowing the conditional probability P(a|b) to be the probability of the logical conditional (a|b). This algebra has been rigorously formulated in [63-67]. A brief review of this development follows.

3.8 Formulation of the Conditional Event Algebra ($\mathcal{B}|\mathcal{B}$)

Start with an initial Boolean algebra, $\mathcal{B}$, of propositions or events such as

1) All events generated by a surveillance track file or files

2) The Boolean algebra generated by any finite or infinite set of propositions
3) The Boolean algebra of subsets of a probability sample space Ω
4) {All 64 subsets of the six-element sample space $\{1, 2, 3, 4, 5, 6\}$}

Boolean set operation "and" is represented by either $\cap$ or $\wedge$. The Boolean operation "or" is represented by $\cup$ or $\vee$. "not" is represented by ($'$) or ($'$).

Let (B | B) denote the set of ordered pairs, {(a|b): a, b in B}, called the set of *conditionals*, "a given b", of B. The proposition or event "b" is called the *condition, premise* or *antecedent* and the proposition or event "a" is called the *consequent* or *conclusion*.

Just as two Boolean propositions or events may be equivalent, that is, refer to the same set of occurrences, so also may two conditionals be equivalent. Analogously, two integer fractions may be equal but look different.

3.9 Equivalent Conditionals
It was Bruno De Finetti in 1936 [90] who first formulated this definition. P. Calabrese 1987 [63] independently rediscovered it:

Two conditional statements (a|b) and (c|d) are equivalent (=) provided: 1) Their conditions, b and d, are equivalent propositions or events, and 2) Their conclusions, a and c, are equivalent when their common condition is true. In symbols,

$$(a|b) = (c|d) \text{ provided that } b = d \text{ and } ab = cd,$$

where juxtaposition, "ab", denotes "a $\wedge$ b", that is, "a and b".

In other words, two conditionals are equivalent when they have equivalent premises and their conclusions are equivalent assuming that common premise.

This equivalence relation on conditionals implies that for all propositions a and b,

$$(a|b) = (ab \mid b)$$

and also that for all $a \in B$, (1|0) = (a|0) = (0|0). (1|0) is the "inapplicable" or "undefined" conditional and is denoted U. The Boolean 1 and 0 propositions are represented by (1|1) and (0|1) respectively.

Thus, in any instance a conditional (a|b) can have any one of 3 truth values:

(a|b) is *true*   if (a|b) = (1|1) = 1, i.e., if a is true and b is true
(a|b) is *false*   if (a|b) = (0|1) = 0, i.e., if a is false and b is true
(a|b) is *inapplicable* if (a|b) = (1|0) = U, i.e., if b is false

Thus, (a|b) is *true* on a ∧ b, *false* on a' ∧ b, and *inapplicable* on b'.

Note also that "not true" means "false or inapplicable"; "not false" means "true or inapplicable"; "not inapplicable" means "true or false". CEA (CPL) provides clear distinctions in terminology. For example, "if a then b" is "not false" on the instances of a ∨ b', but it is "true" on the smaller set of instances of a ∧ b. No such distinction is available in Boolean algebra.

3.10 Operations on Conditionals
Each of the three operations defined below agrees with the corresponding Boolean operation when applied to conditionals with equivalent conditions. Therefore they extend the Boolean operations.

3.10.1 Relative Negation
The relative negation of "a given b" is the "negation of a, given b". That is,

$$(a|b)' = (a' | b),$$

and the latter has probability 1 - P(a|b).

3.10.2 Disjunction
Concerning disjunction, "If b then a, or if d then c" means "If either conditional is applicable then at least one is true". That is,

$$(a|b) \vee (c|d) = (ab \vee cd) | (b \vee d)$$

3.10.3 Conjunction
Concerning conjunction, "If b then a and if d then c" means "if either conditional is applicable then one is true while the other is not false". That is,

$$(a|b) \wedge (c|d) = [\,ab(c \vee d') \vee (a \vee b')cd\,] \,|\, (b \vee d)$$
$$= (abd' \vee abcd \vee b'cd \,|\, b \vee d),$$

which also means "if either conditional is applicable then either they are both true or else one is true while the other is inapplicable." The latter formula also follows by the standard De Morgan formulas in Boolean algebra relating conjunction and negation to disjunction and negation.

The algebra (B | B) of conditionals includes the original Boolean algebra B as those conditionals (B | Ω), where Ω is the universal event. In logical notation these are the conditionals (B | 1) whose condition is certain. Analogously, these are like the integer fractions whose denominators are 1. The proposition a is identified with the conditional (a | 1). Fixing the condition b also yields a Boolean algebra (B | b).

3.10.4 Algebraic Characterization
These operations corresponding to "not", "and" and "or" allow the usual manipulations, although the resulting system is not wholly Boolean. They can be characterized as the only associative, commutative, idempotent operations on conditionals whose two components are Boolean and for which (c|d) $\vee$ (c|d') = (c | d $\vee$ d'). That is, "c given d or c given not d" is equivalent to c without further condition.

3.10.5 Iterated Conditioning
A conditional (c|d) may itself be a condition for another proposition or conditional proposition. Due to the largely unrecognized third truth status of conditional statements, to a great degree, natural language is ambiguous about such iterated conditioning. Without additional qualification the iterated form ((a|b) | (c|d)), "(a|b) given (c|d)", could consistently be taken to mean any one of the following:

"a given b and (c|d)"	- (a | b $\wedge$ (c|d))
"a given b and (c|d) are true"	- (a | b $\wedge$ c $\wedge$ d)
"(a|b) not false given (c|d) is not false"	- (a $\vee$ b' | c $\vee$ d')
"(a|b) true given (c|d) is not false"	- (a $\wedge$ b | c $\vee$ d')
"(a|b) not false given (c|d) is true"	- (a $\vee$ b' | c $\wedge$ d)
"(a|b) true given (c|d) is true"	- (a $\wedge$ b | c $\wedge$ d)

Since it can be shown that $b \wedge (c|d) = b \wedge (c \vee d')$, this means that the first possibility (which will be the default interpretation) reduces to

$$(a|b) | (c|d) = (a | b(c \vee d'))$$

That is, "(a|b) given (c|d)" means "a given that b and (c|d) are not false".

Note that from the above whenever a conditional proposition "if d then c" is itself a condition, then the corresponding (material conditional) proposition, "either c or else not d", can be used in its place (as is commonly done in 2-valued logical proof arguments. The conditional (c|d) also acts like its corresponding material conditional, $(c \vee d')$, when conjoined ($\wedge$) with any (otherwise unconditioned) proposition b.

3.11 Operations on Conditionals - Further Motivations

Although intuitive and technical motivations for these operations have been published several times (see for instance [68]), there are some people who still consider them to be debatable. Therefore a restatement and refinement of those motivations is appropriate here. This can also be done using indicator functions as adopted and employed by Walker [82] to list all possible candidates for operations on conditionals.

3.11.1 Conditionals and Indicator Functions

As previously demonstrated (Also see for instance [60, p. 334] or [63, pp. 234-235]) any conditional (a|b) can be represented as a domain-restricted, measurable indicator function defined as follows:

$$(a|b)(\omega) = \begin{cases} 1, & \omega \text{ in } ab \\ 0, & \omega \text{ in } a'b \\ U, & \omega \text{ in } b' \end{cases}$$

U means "undefined". Conversely, any such indicator function assigning 1 and 0, respectively, to disjoint events ab and a'b, and which is undefined elsewhere, determines a unique conditional (a|b).

3.11.2 Boolean Functions
By a Boolean function is meant a polynomial built up from the identity function and constant functions on events using negation, conjunction and disjunction a finite number of times.

By the well-known Fundamental Theorem of Boolean algebra such Boolean functions are completely determined by their values, $f(1)$ and $f(0)$, for the two Boolean values 1 and 0. In fact, a Boolean function f of one Boolean variable x is always of the form

$$f(x) = (f(1) \wedge x) \vee [f(0) \wedge x')$$

and a Boolean function f of two variables x and y is always of the form

$$f(x, y) = f(1,1)xy \vee f(1,0)xy' \vee f(0,1)x'y \vee f(0,0)x'y',$$

where for the sake of readability juxtaposition has replaced the conjunction operator in the latter formula.

3.11.3 Operations on Conditionals
We restrict attention to operations on conditionals that are defined in such a way that the two components of the image conditional are Boolean functions of the component events of the operands. For instance, the negation operation (') defined on conditionals in Section 3.10.1 is of the form

$$(a|b)' = (g(a,b) | h(a,b))$$

where $g(a, b)$ and $h(a, b)$ are Boolean functions of a and b. The disjunction operation, $\vee$, is of the form

$$(a|b) \vee (c|d) = (g(a,b,c,d) | h(a,b,c,d))$$

where g and h are Boolean functions of a, b, c, and d.

Now it is easy to see that any such operation f on conditionals into the set $(\mathcal{B}|\mathcal{B})$ of conditionals determines a unique three-valued truth table by simply setting a and b in turn to the values 1 or 0. If the operation is a one-place function like the negation operation, then the truth table will be:

One-Place Truth Table on Conditional Propositions Generated by f:

(x|y)	f(x|y)
(1|1)	(g(1,1) | h(1,1))
(0|1)	(g(0,1) | h(0,1))
(0|0)	(g(0,0) | h(0,0))

It is also clear that conversely, any such truth table function k determines a unique operation on conditionals. A one-place truth table function k generates a one-place conditional operation as follows:

$$[k(a\,|\,b)](\omega) = k((a\,|\,b)(\omega)) = \begin{cases} k(1), \text{ if } \omega \in ab, \\ k(0), \text{ if } \omega \in a'b, \\ k(U), \text{ if } \omega \in b'. \end{cases}$$

That is, k assigns the measurable indicator function $k((a|b)(\omega))$ to any three-valued, measurable indicator function $(a|b)(\omega)$.

A similar statement holds for a two-place, three-valued truth table. A two-place operation like disjunction has a truth table k of the form:

Two-place Truth Table k

k	1	0	U
1	k(1, 1)	k(1, 0)	k(1, U)
0	k(0, 1)	k(0, 0)	k(0, U)
U	k(U, 1)	k(U, 0)	k(U, U)

Conversely, such a truth table k defines a unique two-place operation f on conditionals by defining its associated indicator function f as follows:

$$f((a|b), (c|d))(\omega) = k((a|b)(\omega), (c|d)(\omega))$$

Summarizing the four operations on conditionals in logical terminology:

	AND			OR			GIVEN			NOT
	T	F	U	T	F	U	T	F	U	
T	T	F	T	T	T	T	T	U	T	F
F	F	F	F	T	F	F	F	U	F	T
U	T	F	U	T	F	U	U	U	U	U

3.11.4 Motivations for the "not", "and" and "or" Operations

It has been shown above that the operations on conditionals are each characterized by a three-valued truth table. The negation operation (') has a truth table of the form

Negation Truth Table on Conditional Propositions
Not (')

1	x
0	y
U	z

In this table $1 = (1|1)$, $0 = (0|1)$ and $U = (0|0)$. Since negation of conditionals is intended to extend the Boolean negation operation, the values x and y must be 0 and 1, respectively. So only z is free. But since the double negation of a conditional should be the conditional back again, z must be U. (If $z = 1$, then $(U')' = 1' = 0 \neq U$; and if $z = 0$, then $(U')' = 0' = 1 \neq U$.) Thus the negation operation can only have the truth table above with $x = 0$, $y = 1$ and $z = U$. So $(a|b)'$ must be $(a'|b)$ since the latter conditional has the same truth table.

Similarly, the conjunction operation ($\wedge$) has a truth table of the form:

Two-place Truth Table for Conjunction ($\wedge$)

$\wedge$	1	0	U
1	1	0	x
0	0	0	y
U	x	y	z

Again, since conjunction of conditionals is intended to extend conjunction of Boolean events, 1's and 0's have been inserted in the table in the appropriate places leaving only five entries undetermined as designated by unknowns x, y and z.

Since conjunction is intended to be commutative the table must be symmetric about the diagonal. Thus again only three values are still free. Since conjunction should be idempotent, it follows that z = U, and so there are only four possible ways to finish the table.

To motivate the choice of values for x and y, consider how we normally prove a statement A in case B is true or in case its negation B' is true. We can show that A is true in case B is true and that A is true in case B' is true, and so therefore prove that A is true. That is, to show A is true, we can show that A is true given B is true and that A is true given that B' is true. That is, we use that

$$(A|B) \wedge (A|B') = A$$

(We also know that if A is true, then (A|B) and (A|B') will each be true or inapplicable.) Setting B = 0 and A = 1 in the above equation yields that

$$(1|0) \wedge (1|1) = 1.$$

That is, U ∧ 1 = 1. Therefore x = 1 in the conjunction table.

Similarly, we can show that A is impossible (0) by showing that both (A|B) is false and (A|B') is false. That is, we use that

$$(0|0) \wedge (0|1) = 0.$$

This also follows by simply setting A = 0 instead of A = 1 above. So U ∧ 0 = 0, and therefore in the conjunction table y must be 0. That completes the conjunction truth table.

(Alternately, consider how a questionnaire with conditional questions is interpreted when some of the conditional questions do not apply to an individual: The answers to questions are basically conjoined and any inapplicable questions are ignored. They do not make the whole set of answers inapplicable or undefined.)

The disjunction operation (∨) has a truth table like

Two-place Truth Table for Disjunction (v)

v	1	0	U
1	1	1	x
0	1	0	y
U	x	y	U

and like the conjunction operation most of the table is determined because the operation is intended to extend Boolean disjunction to conditionals and be commutative and idempotent. These properties will all be satisfied and the table completed by simply specifying that the De Morgan's laws should also hold for conditionals. That is, we require that the negation of a conjunction of conditionals be equivalent to the disjunction of the negations of those conditionals. So since (0|0) ∧ (0|1) = 0, taking negations on both sides, yields that (0|0)' v (0|1)' = 0', which is equivalent to (1|0) v (1|1) = 1. That is, U v 1 = 1. So in the disjunction table x must be 1. Similarly, since (1|0) ∧ (1|1) = 1, taking negations on both sides yields that U v 0 = 0. So in the table y must be 0. We can now summarize the results of this subsection in the following theorem.

3.11.5 Algebraic Characterization Theorem

The only unary operation ' on conditionals whose double operation is idempotent and that extends Boolean negation on events and whose components are Boolean functions of the components of the operand conditional is that of Section 3.10.1, namely, (a|b)' = (a'|b). Furthermore, the only commutative, idempotent binary operations on conditionals that extend the conjunction and disjunction operations on events and whose components are Boolean functions of the components of the operands, and which satisfy the De Morgan's formulas, and which satisfy the property (c|d) ∧ (c|d') = (c|d v d') for all events c and d, are those of subsections 3.10.2 and 3.10.3, namely, (a|b) v (c|d) = (ab v cd | b v d) and (a|b) ∧ (c|d) = (abd' v abcd v b'cd | b v d).

3.11.6 Alternate Formulations

Adams [2] also defines what he calls "quasi-operations" on conditionals that are equivalent to those of Sections 3.7.2 and 3.7.3 but he does not settle on an iterated conditional operation even though he does identify situations when update information is conditional. Adams refers to his operations as "quasi" specifically because they are not monotonic, that is, for example, the conjunction of conditionals does not necessarily imply each of the conditionals being conjoined. (But that is just one way operations on conditionals differ from operations on ordinary events.)

Schay [60] defines two 3-operation algebras, the first of which Goodman, Nguyen and Walker erroneously identify as equivalent to the author's first three operations (See p. 6 and p. 92 of [75]). Then this error was propagated by Hailperin [85, p. 266]. But the truth is that the author's disjunction operation is in one of Schay's systems and the conjunction operation is in the other of Schay's systems, and so neither of Schay's systems contains the three operations of Sections 3.10.1-3.10.3. No wonder Hailperin, as he says [85, p. 261], gets results so different from those of Schay.

Concerning the third truth-value, Hailperin quotes Schay, who says for conditionals a and b: "... if a is defined and b is undefined, then we put max{a,b} = a and min{a,b} = a". Hailperin [85, p. 262] finds it "difficult to conceive of a rationale for defining max and min in this manner". The difficulty is in the misinterpretation of the third truth-value. When a conditional is "undefined" that is nothing like saying that its truth-value is unknown. In the latter situation assigning a truth-value or better, a probability, between 0 and 1 is appropriate. But a conditional with a false condition is not somewhat true; it does not deserve to have a truth-value between 0 and 1 as though it were somewhat true. It is simply inapplicable – a completely different category. Conjoining a true or false statement with one that is inapplicable leaves the applicable statement unchanged. Is that so difficult to conceive? Is that not what we do when we skip an inapplicable question on a questionnaire and fill out the other questions? In principle, we expect the reader to conjoin all of our answers, conditional or not, and ignore the inapplicable questions. We do not expect the reader to declare the whole form "undefined" merely because one inapplicable conditional was encountered.

Hailperin also takes issue with this author [85, p. 264] for claiming that when rolling one 6-sided die the statement "if the roll is even, then it will be a 6 or if the roll is odd, then it will be a 5" is intuitively equivalent to the statement "The roll will be a 6 or a 5". He does not see it as so clear cut and apparently finds the Goodman-Nguyen alternative translation as "If the roll is 6 or 5, then the roll will be 6 or 5", to be intuitively preferable, even though it has a conditional probability of 1.

Shifting to the conjunction operation in keeping with the motivations presented in this book, one wonders whether Hailperin finds the statement "If the roll is even, then it will be a 5 or a 6 and if the roll is odd, then it will be a 5 or a 6" to be intuitively equivalent to "the roll will be a 5 or a 6 whether or not it is even". That is, after all, the intuitive meaning of a proof by cases, and it also works when the conclusion is uncertain. But the conjunction operation favored by Hailperin, Goodman and Nguyen has it that the statement is instead equivalent to "if the roll is 1, 2, 3, or 4, then falsity". That might be an acceptable translation for some logical purposes but it has conditional probably 0 instead of 2/6, which is the intuitive probability of getting a 5 or 6 in one roll of a die given that the roll is odd or even.

3.11.7 Motivations for the Iterated Conditioning Operation

The fourth operation on conditionals, defined in Section 3.7.4, is essential for closure of the algebraic system. Without it, conditioning cannot be performed on conditionals, and deduction or inference of a conditional by another conditional remains out of reach of the syntax of the algebra. The ability of the system to express other operations on conditionals is also greatly expanded by the inclusion of the fourth operation. As shown by Tyszkiewicz [86] and his co-authors, the other operations on conditionals defined by Goodman et al. [75] can all be expressed in terms of the four operations defined in Sections 3.7.1-3.7.4.

When discussing the author's system of conditionals some authors [88, p. 1706], have ignored the iterated conditioning operation as though it were a thing apart, preferring to characterize the system in terms of just the first three operations and identifying it with the first three operations of the system of Sobocinski [46, 89]. While the first three operations of Sobocinski are in fact equivalent to those of Sections 3.7.1-3.7.3, the fourth is quite different from Sobocinski's implication operation.

When the fourth operation is included it is clear that the operations of Sections 3.7.1-3.7.4 are not a repetition of those that have previously been explored in the literature. (This is also apparent when it is noted that an iterated conditional ((a|b) | (c|d)) is not interpreted to be an implication (c|d) $\Rightarrow$ (a|b), as is done by most authors. Nor is the conditional itself an implication. Rather, implications have been separately defined (See Calabrese [69] and Chapter 6.)

Now the most straightforward way to motivate the iterated conditional operation of Section 3.7.4 is to extend the following rule of iterated conditioning:

$$((a|b) \mid c) = (a \mid b \wedge c)$$

Here again, mathematicians routinely prove theorems by successive conditioning according to the above formula. We very often read arguments of the form "if c is true, then if also b is true, then a will be true". The proof will then proceed to show that if both b and c are true, then a will be true, and no one will dispute the matter. In this regard, Adams [2, p. 33] mentions this iterated conditional simplification as intuitively plausible but says that it would entail "giving up modus ponens in application to conditionals with conditional consequents". Adams' example is that ((A|B) | A) would be certain and so we should always infer (A|B) from A, but, he says, we know independently that this inference is not always sound. However there seems to be no real problem with this inference when we are talking about conditionals instead of implications. When A is true the conditional (A|B) will be true as long as B is true, or inapplicable if B is false. We are not inferring that (A|B) is always true.

The general iterated conditional ((a|b)|(c|d)) can then be reduced using this formula as follows:

$$((a|b) \mid (c|d)) = (a \mid (b \wedge (c|d))),$$

which can be further reduced to (a | (b ∧ (c ∨ d'))), which is a conditional with Boolean components as required.

The truth table for this conditioning operation, (|), is:

Two-place Truth Table for the Conditional Operation (|)

(|)	1	0	U
1	1	U	1
0	0	U	0
U	U	U	U

where the first column and top row are the input values (1, 0 or U) of the table, the top row providing the "condition" and the first column providing the "consequent". If the condition = 0 then the results is U no matter what the consequent. The last column expresses the interpretation that an undefined condition leaves the consequent unchanged. That is, (1|U) = 1, (0|U) = 0 and (U|U) = U.

Adding to the premise of the characterization theorem of Section 3.11.5 the iterated conditioning rule

$$((a|b) \mid (c|d)) = (a \mid (b \wedge (c|d))),$$

for any events a, b, c, and d the whole system of four operations on conditionals (Sections 3.7.1-3.7.4) has been algebraically characterized.

3.12 Minimal Axioms for Operations in CPL

Having defined conditional propositions and conditional events generated by a probability space $\mathcal{P}$, minimal assumptions for CPL can now be identified. It is not known whether the 5 axioms of the following subsection are axiomatically independent. To show axiomatic independence of an axiom X from the other four axioms, a specific algebra $\mathcal{L}(X)$ of conditionals needs to be exhibited that does not satisfy axiom X, but which does satisfy the other four axioms. If such an algebra $\mathcal{L}(X)$ exists for each of the 5 axioms, then the system of 5 axioms is said to be a system of independent axioms.

3.13 Theorem Characterizing CPL

Let a, b, c, d and e be arbitrary propositions and (e|f) any conditional proposition. The following axioms are sufficient to generate CPL.

1) The Boolean algebra operations ($\wedge$, $\vee$, and $'$) are extend to conditional propositions of the form (a | 1).

2) Law of Proof by Cases: (a|b) $\wedge$ (a|b$'$) = a

3) One de Morgan Law for Conditionals: [(a|b) $\wedge$ (c|d)]$'$ = (a|b)$'$ $\vee$ (c|d)$'$

4) Law of Iterated Conditioning: ((a|b) | (e|f)) = ((a | b $\wedge$ (e|f))

5) The double negation operation is idempotent, and the operations of ∧ and ∨ on conditionals are idempotent and communicative.

Proof of Theorem 3.13 The negation operation (′) has been shown to be (a|b)′ = (a′|b) since it is a 1-place function on conditionals extending negation and therefore determined by it values for the three conditionals {(1|1), (0|1) and (0|0)}. But the first two conditionals are the extensions of 1 and 0. So they are each other's negation. Therefore the negation of (0|0) must be (0|0) in order that its double negation will again be (0|0).

Concerning disjunction, replacing a with a′ in 2) and applying the de Morgan formula yields that (a|b) ∨ (a|b′) = a. Furthermore, if bc = 0, then (a | b ∨ c) = (a|b) ∧ (a|c) because using 2) and 4)

$$\begin{aligned}(a \mid b \vee c) &= [(a|b) \wedge (a|b')] \mid (b \vee c) \\ &= [(a|b) \mid (b \vee c)] \wedge [(a|b') \mid (b \vee c)] \\ &= [a \mid b(b \vee c)] \wedge [a \mid b'(b \vee c)] \\ &= (a|b) \wedge (a \mid b'c) \\ &= (a|b) \wedge (a|c)\end{aligned}$$

using that b′c = c since bc = 0. And again using de Morgan, it follows that if bd = 0, then also

$$(a \mid b \vee c) = (a|b) \vee (a|c)$$

The general disjunction operation can now be derived:

$$\begin{aligned}(a|b) \vee (c|d) &= (ab|b) \vee (cd|d) \\ &= (ab \mid bd' \vee bd) \vee (cd \mid bd \vee b'd) \\ &= (ab \mid bd') \vee (ab \mid bd) \vee (cd \mid bd) \vee (cd \mid b'd) \\ &= (ab \mid bd') \vee (ab \vee cd \mid bd)) \vee (cd \mid b'd) \\ &= (ab \vee cd \mid bd') \vee (ab \vee cd \mid bd)) \vee (ab \vee cd \mid b'd) \\ &= (ab \vee cd \mid bd' \vee bd \vee b'd) \\ &= (ab \vee cd \mid b \vee d)\end{aligned}$$

where (ab | bd′) = (ab ∨ cd | bd′) because (cd | bd′) = (0 | bd′), and similarly for (ab ∨ cd | b′d).

The general conjunction operation then follows from de Morgan:

$$(a|b) \wedge (c|d) = [(a|b)' \vee (c|d)']'$$
$$= [(a'|b) \vee (c'|d)]'$$
$$= [(a'b \vee c'd \mid b \vee d)]'$$
$$= [(a'b \vee c'd)' \mid b \vee d]$$
$$= [(a \vee b') \wedge (c \vee d') \mid b \vee d]$$
$$= [(ab \vee b')(cd \vee d') \mid b \vee d]$$
$$= [abd' \vee abcd \vee b'cd \vee b'd' \mid b \vee d]$$
$$= (abd' \vee abcd \vee b'cd \mid b \vee d)$$

Alternately, and instructively, the conjunction operation can be derived directly by expanding as follows:

$$(a|b) \wedge (c|d) = (ab \mid b) \wedge (cd \mid d)$$
$$= (ab \mid bd' \vee bd) \wedge (cd \mid bd \vee b'd)$$
$$= (ab \mid bd') \wedge (ab \mid bd) \wedge (cd \mid bd) \wedge (cd \mid b'd)$$
$$= (ab \mid bd') \wedge (ab \wedge cd \mid bd) \wedge (cd \mid b'd)$$

But $(ab \mid bd') = (abd' \mid bd') = (abd' \vee abcd \vee cdb' \mid bd')$ = because abcd and cdb' are 0 given bd', and therefore add nothing to $(ab \mid bd')$. Similarly, $(cd \mid b'd) = (abd' \vee abcd \vee cdb' \mid b'd)$, and $(ab \wedge cd \mid bd) = (abd' \vee abcd \vee cdb' \mid bd)$. Substituting for these yields

$$(a|b) \wedge (c|d) = (abd' \vee abcd \vee cdb' \mid bd')$$
$$\vee (abd' \vee abcd \vee cdb' \mid bd)$$
$$\vee (abd' \vee abcd \vee cdb' \mid b'd)$$
$$= (abd' \vee abcd \vee cdb' \mid bd' \vee bd \vee b'd)$$
$$= (abd' \vee abcd \vee cdb' \mid b \vee d)$$

Corollary: In the context of Theorem 3.13, then

$$((a|b) \wedge (c|d) \mid e) = ((a|b) \mid e) \wedge ((c|d) \mid e)$$

Proof of Corollary:

$$((a|b) | e) \wedge ((c|d) | e) = (a|be) \wedge (c|de)$$
$$= (a(be)(de)' \vee a(be)c(de) \vee (be)'c(de) | (be \vee de))$$
$$= (a(be)(de)' \vee a(be)c(de) \vee (be)'c(de) | (b \vee d)e)$$
$$= ((a(be)d' \vee abcde \vee b'c(de) | (b \vee d)e)$$
$$= ((abd' \vee abcd \vee b'cd | b \vee d) | e)$$
$$= ((a|b) \wedge (c|d) | e)$$

3.14 Fundamental Theorem of Boolean Algebra Extended to Conditionals

Here is an extension of the Fundamental Theorem of Boolean algebra to conditional propositions. Recall (see, for instance, P. Rosenbloom [49], p. 5) that a Boolean function of one variable is a function that can be formed by starting with constant functions and the identity function and applying to these the operations of conjunction ($\wedge$), disjunction ($\vee$), and negation ($\neg$). This definition can easily be extended to functions of any finite number of variables. For Boolean functions f of one variable x the fundamental theorem of Boolean algebra states that

$$f(x) = (f(1) \wedge x) \vee (f(0) \wedge x'),$$

where x' has been written for $\neg x$ to condense notation. It is then a corollary that any such Boolean function is completely determined by its action on the two propositions 1 and 0. For Boolean functions of two variables, this becomes

$$f(x,y) = f(1,1)xy \vee f(1,0)xy' \vee f(0,1)x'y \vee f(0,0)x'y'$$

where juxtaposition has replaced $\wedge$. The above functions are said to be expressed in "disjunctive normal form".

3.14.1 Definition of Conditional Boolean Functions

A conditional Boolean function of one conditional variable (x|y) is a function f: $L/L \to L/L$ given by

$$f(x|y) = (g(x,y) | h(x,y))$$

where g and h are Boolean functions of two Boolean propositional variables x and y. Similarly, a conditional Boolean function of two conditional variables is a function f given by

$$f((x|y), (w|z)) = (g(x,y,w,z) | h(x,y,w,z))$$

where g and h are Boolean functions of four Boolean propositional variables. f is a conditional Boolean function of n conditional variables if

$$f(x_1|y_1, x_2|y_2 \ldots x_n|y_n) = (w|z)$$

where $w = g(x_1, y_1, x_2, y_2, \ldots, x_n, y_n)$ and $z = h(x_1, y_1, x_2, y_2, \ldots, x_n, y_n)$ are Boolean functions of 2n variables.

3.14.2 Fundamental Theorem of Boolean Algebra Extended to Conditionals

If f is a conditional Boolean function of one variable then f can be uniquely expressed by

$$f(x|y) = (f(1|1) | xy) \vee (f(0|1) | x'y) \vee (f(0|0) | y')$$

This expression for f(x|y) will be called the disjunctive normal form.

Lemma 1 for Theorem 3.14.2 If a, b, c, and d are any four propositions then

$$(a \vee b) | (c \vee d) = (a|c) \vee (a|d) \vee (b|c) \vee (b|d)$$

Proof of Lemma 1

$$\begin{aligned}(a \vee b) | (c \vee d) &= (a | (c \vee d)) \vee (b | (c \vee d)) \\ &= (a|c) \vee (a|d) \vee (b|c) \vee (b|d)\end{aligned}$$

Notice that the conditions c and d need not be disjoint for the second equality to hold. This lemma can easily be extended to any two finite sets of propositions $\{a_i: i = 1, 2 \ldots n\}$ and $\{b_j: i = 1, 2 \ldots m\}$:

$$(a_1 \vee a_2 \vee \ldots \vee a_n) | (b_1 \vee b_2 \vee \ldots \vee b_m) = \bigvee_{i,j} (a_i | b_j)$$

Proof of Fundamental Theorem 3.14.2 Since f must be a well-defined function, and since $(x|y) = (xy|y)$, it follows that $f(x|y) = f(xy|y) = g(xy,y) \mid h(xy,y)$. Now, applying the fundamental theorem of Boolean algebra to g yields

$$g(xy,y) = g(1,1)(xy)y \lor g(1,0)(xy)y' \lor g(0,1)(xy)'y \lor g(0,0)(xy)'y'$$
$$= g(1,1)xy \lor g(1,0)(0) \lor g(0,1)x'y \lor g(0,0)y'$$
$$= g(1,1)xy \lor g(0,1)x'y \lor g(0,0)y'.$$

Similarly, $h(xy,y) = h(1,1)xy \lor h(0,1)x'y \lor h(0,0)y'$. Therefore, using the lemma

$f(x|y)$
$= [g(1,1)xy \lor g(0,1)x'y \lor g(0,0)y'] \mid [h(1,1)xy \lor h(0,1)x'y \lor h(0,0)y']$
$= [g(1,1)xy \mid h(1,1)xy] \lor [g(1,1)xy \mid h(0,1)x'y] \lor [g(1,1)xy \mid h(0,0)y']$
$\quad \lor [g(0,1)x'y \mid h(1,1)xy] \lor [g(0,1)x'y \mid h(0,1)x'y]$
$\quad \lor [g(0,1)x'y \mid h(0,0)y'] \lor [g(0,0)y' \mid h(1,1)xy]$
$\quad \lor [g(0,0)y' \mid h(0,1)x'y] \lor [g(0,0)y' \mid h(0,0)y'].$

Now only one of the three conditionals with condition $h(1,1)xy$ is non-zero. The others are equivalent to $[0 \mid h(1,1)xy]$ because both $(x'y)(xy) = 0$ and $(y')(xy) = 0$. Similarly for the conditionals with conditions $h(0,1)x'y$ and $h(0,0)y'$. Thus

$f(x|y)$
$= [g(1,1)xy \mid h(1,1)xy] \lor [g(0,1)x'y \mid h(0,1)x'y] \lor [g(0,0)y' \mid h(0,0)y']$
$= [g(1,1) \mid h(1,1)xy] \lor [g(0,1) \mid h(0,1)x'y] \lor [g(0,0) \mid h(0,0)y']$
$= [(g(1,1) \mid h(1,1)) \mid xy] \lor [(g(0,1) \mid h(0,1)) \mid x'y] \lor [(g(0,0) \mid h(0,0)) \mid y']$
$= [f(1|1) \mid xy] \lor [f(0|1) \mid x'y] \lor [f(0|0) \mid y'].$

To show uniqueness, suppose f and j are two conditional Boolean functions of $(x|y)$. Then f and j can be expressed by

$f(x|y) =$
$[g(1,1)xy \lor g(0,1)x'y \lor g(0,0)y'] \mid [h(1,1)xy \lor h(0,1)x'y \lor h(0,0)y']$

$j(x|y) =$
$[k(1,1)xy \lor k(0,1)x'y \lor k(0,0)y'] \mid [m(1,1)xy \lor m(0,1)x'y \lor m(0,0)y']$

where g, h, k and m are Boolean functions. Now if the right-hand sides of the above equations are equal, then by the definition of equivalent conditionals,

$$h(1,1)xy \lor h(0,1)x'y \lor h(0,0)y'$$
$$= m(1,1)xy \lor m(0,1)x'y \lor m(0,0)y'$$

and

$$g(1,1)h(1,1)xy \lor g(0,1)h(0,1)x'y \lor g(0,0)h(0,0)y'$$
$$= k(1,1)m(1,1)xy \lor k(0,1)m(0,1)x'y \lor k(0,0)m(0,0)y'$$

Clearly, from the first equation $h(1,1) = m(1,1)$, $h(0,1) = m(0,1)$, and $h(0,0) = m(0,0)$. Therefore the second equation becomes

$$g(1,1)h(1,1)xy \lor g(0,1)h(0,1)x'y \lor g(0,0)h(0,0)y'$$
$$= k(1,1)h(1,1)xy \lor k(0,1)h(0,1)x'y \lor k(0,0)h(0,0)y'$$

Furthermore, using the uniqueness for the Boolean case of the Fundamental Theorem or directly,

$$g(1,1)h(1,1) = k(1,1)h(1,1)$$
$$g(0,1)h(0,1) = k(0,1)h(0,1)$$
$$g(0,0)h(0,0) = k(0,0)h(0,0).$$

Therefore, $[g(1,1) \mid h(1,1)] = [k(1,1) \mid m(1,1)]$, and so $f(1|1) = j(1|1)$. Similarly, $f(0|1) = j(0|1)$ and $f(0|0) = j(0|0)$. Therefore

$$f(x|y) = (f(1|1) \mid xy) \lor (f(0|1) \mid x'y) \lor (f(0|0) \mid y').$$
$$= (j(1|1) \mid xy) \lor (j(0|1) \mid x'y) \lor (j(0|0) \mid y') = j(x|y)$$

This completes the proof of the Fundamental Theorem.

The extension to conditional Boolean functions of two or more variables is straightforward but messy to write. For a function f of two conditional variables there are nine disjoined terms. Writing $f((x|y), (w|z))$ as $f(x|y, w|z)$ to reduce parentheses the result is

$f(x|y, w|z) =$
   $[f(1|1, 1|1) \mid xywz] \lor [f(1|1, 0|1) \mid xyw'z] \lor [f(1|1, 0|0) \mid xyz']$

$\vee$ [f(0|1, 1|1) | x'ywz] $\vee$ [f(0|1, 0|1) | x'yw'z] $\vee$ [f(0|1, 0|0) | x'yz']
$\vee$ [f(0|0, 1|1) | y'wz] $\vee$ [f(0|0, 0|1) | y'w'z] $\vee$ [f(0|0, 0|0) | y'z']

It is now possible to prove Goodman's representation theorem as a corollary to this extended fundamental theorem, where again for simplicity, consider conditional Boolean functions of just one variable.

Corollary 1 of Fundamental Theorem A conditional Boolean function f of one variable induces a unique 3-valued, 2-place truth table on L/L. Conversely, each such 3-valued, 2-place truth table induces a unique conditional Boolean function of one variable.

Corollary 2 of Fundamental Theorem There are 3^3 different conditional Boolean functions of one variable; there are 3^{3^n} different conditional Boolean functions of n variables.

Proof of Corollary 1 Given a conditional Boolean function f of one variable define the truth table function t by t(x|y) = f(x|y) for (x|y) $\in$ {(1|1), (0|1), (0|0)}. Conversely, if t(x|y) is a 3-valued truth table function with domain values (1|1), (0|1), or (0|0), then define the conditional Boolean function f by:

f(x|y) = [t(1|1) | xy] $\vee$ [t(0|1) | x'y] $\vee$ [t(0|0) | y']

By the uniqueness part of the fundamental theorem, different truth table functions t induce different conditional Boolean functions f.

Proof of Corollary 2 By the fundamental theorem, f can be uniquely expressed in disjunctive normal form and so there are three possible assignments to each of the three domain elements {(1|1), (0|1), (0|0)}. Thus there are 3^3 different possible assignments and that number of possible conditional Boolean functions of one variable. For conditional Boolean functions of two variables there are 3^2 different domain elements each of which can be assigned any one of 3 different values. So there are 3^{3^2} different possible assignments and that number of different conditional Boolean functions of two variables. The generalization to n values is straightforward.

Chapter 4
Operating on Functions with Variable Domains

> By using indicator functions... the difficulty of A | B not being an object is eliminated, for A | B is just the indicator function of the set A restricted to the set B, that is, (A | B) is a partial function whose domain is B.... The use of such partial functions requires care in formulating the algebra of functions in which we are interested, for functional addition X | A + Y | B will not be defined when A ≠ B but A ∩ B ≠ ∅.
>
> P. Suppes & M. Zanotti [92, Ch.2, p.10, first publ. 1982]

4.1 Synopsis

The sum, difference, product and quotient of two functions with different domains are usually defined only on their common domain. This chapter extends these definitions so that the sum and other operations are defined anywhere that at least one of the components is defined. This idea is applied to propositions and events, expressed as indicator functions, to define *conditional* propositions and *conditional* events as three-valued indicator functions that are undefined when their condition is false. Extended operations of "and", "or", "not" and "conditioning" are then defined on these conditional events with variable conditions. The probabilities of the disjunction (or) and of the conjunction (and) of two conditionals are expressed in terms of the conditional probabilities of the component conditionals. In a special case, these are shown to be weighted averages of the component conditional probabilities where the weights are the relative probabilities of the various conditions. Next, conditional random variables are defined to be random variables X whose domain has been restricted by a condition on a second random variable Y. The extended sum, difference, product and conditioning operations on functions are then applied to these conditional random variables. The expectation of a random variable and the conditional expectation of a conditional random variable are recounted. Theorem 4.13 generalizes the standard result that the conditional expectation of the sum of two conditional random variables with disjoint and exhaustive conditions is a weighted sum of the conditional expectations of the component conditional random variables. Because of the extended operations, the theorem is true for arbitrary conditions. Theorem 4.14 gives a formula for the expectation of the product of two conditional random variables. After the definition of independence of two random variables is extended to accommodate the extended operations, it is applied to the formula of Theorem 4.14 to simplify the expectation of a product of conditional random variables. Two examples end the chapter. The first concerns a work force of n workers of different output levels and work shifts. The second example involves two radars with overlapping surveillance regions and different detection error rates. One radar's error rate is assumed to be sensitive to fog and the other radar's error rate is assumed to be sensitive to air traffic density. The combined error rate over the combined surveillance area given heavy fog and moderate air traffic is computed.

4.2 Introduction

From elementary mathematics, we are all familiar with the definitions of the operations of addition, subtraction, multiplication, and division for real-valued functions defined on a common domain D. For each domain element x, the sum function, (f + g), is simply assigned the value (f + g)(x) = f(x) + g(x), the sum of the values of f and g at x, and similarly for the other operations. However, function division, (f/g), requires an extra condition, namely that g(x) not be zero, so that the division can be performed. So (f/g) is said to be "undefined" for any domain values x for which g(x) = 0. Thus already the division operation on functions generates new functions that have *restricted* domains, and in general such divisions will generate functions having *different* domains of definition. This leads to the standard definition of operations on functions whose domains are different: If f and g are defined on D and E respectively, then the sum function (f + g) is defined on the intersection of D and E as follows:

$$(f + g)(x) = \begin{cases} f(x) + g(x) & \text{if } x \in D \cap E, \\ \text{Undefined} & \text{if } x \notin D \cap E \end{cases} \quad (4.1)$$

The difference, product and division functions f − g, f ∗ g, and f / g are similarly defined when D and E are the domains of f and g respectively, but again the quotient (f/g) is also undefined on any zeros of g. Since a summing of f and g cannot be performed for a given domain value x unless both f and g are defined at x, this has seemed to be a reasonable definition, and there has been no reason offered to do it in any other way.

4.3 Extended Operations of Sum, Difference and Product on Real-Valued Functions

However, developments in conditional event algebra [63-67] suggest that there is good reason for expanding the domain of the sum function to include all values of x that are in at least one of the two domains. Using the set theory notation D′ to denote the complement of D, the definition of the sum function (f + g) can advantageously be extended to:

$$(f+g)(x) = \begin{cases} f(x) + g(x) & \text{if } x \in D \cap E \\ f(x) & \text{if } x \in D \cap E' \\ g(x) & \text{if } x \in D' \cap E \\ \text{Undefined} & \text{if } x \in D' \cap E' \end{cases} \quad (4.2)$$

In other words, here the sum function takes the value of f(x) if g(x) is undefined, and takes the value g(x) if f(x) is undefined. It then agrees with the old definition on the restricted domain D ∩ E and is undefined only on the region outside of D ∪ E. The other operations on functions, (f − g), (f ∗ g) and (f/g) can be similarly defined. The product (f ∗ g) is completely analogous to the sum with ∗ in place of +. The difference is:

$$(f-g)(x) = \begin{cases} f(x) - g(x) & \text{if } x \in D \cap E \\ f(x) & \text{if } x \in D \cap E' \\ -g(x) & \text{if } x \in D' \cap E \\ \text{Undefined} & \text{if } x \in D' \cap E' \end{cases} \quad (4.3)$$

The quotient is analogous to the difference:

$$(f/g)(x) = \begin{cases} f(x)/g(x) & \text{if } x \in D \cap E \text{ and } g(x) \neq 0, \\ f(x) & \text{if } x \in D \cap E', \\ (1/g)(x) & \text{if } x \in D' \cap E \text{ and } g(x) \neq 0, \\ \text{Undefined} & \text{if } x \in D' \cap E' \text{ or } g(x) = 0 \end{cases} \quad (4.4)$$

Note that although it is possible in the sum case, for example, to redefine the two functions f and g to be zero instead of undefined and thereby eliminate the need for the extended operations, a subsequent desire to take the product instead of the sum would require another redefinition. Other advantages are exhibited below.

4.4 Propositions, Events and Indicator Functions

These kinds of extended definitions have been shown [63, 66-67] to be useful when defining Boolean-like operations on uncertain conditional propositions or conditional events whose conditions are different.

In a similar vein restricted indicator functions, and their closure under finite addition, have been used successfully by Suppes and Zanotti ([92] p.10 or 91] to define a qualitative relation between pairs of events characterizing the conditional probabilities and conditional expectations of such pairs of events and allowing comparison of conditional probabilities or conditional expectations even when the conditions on the events are different.

For Boolean propositions or events A and B we have familiar and standard operations of "and" ($\wedge$), "or" ($\vee$), and "not" ($\neg$) corresponding to multiplication ($*$), summation ($+$), and negation ($-$) respectively, and also corresponding to intersection ($\cap$), union ($\cup$), and complement ($'$) in the event interpretation. There has been no standard definition of division of propositions or events but now "conditioning" has been recognized to be division. See, for instance, [68].

A proposition A can be represented as a measurable indicator function f_A defined on the universe Ω and taking the value 1 for ω in A and 0 for ω in A':

$$f_A(\omega) = \begin{cases} 1, & \text{if } \omega \in A, \\ 0, & \text{if } \omega \in A' \end{cases} \quad (4.5)$$

With this representation, the standard operations on propositions or events can be expressed in terms of function operations. For instance the negation (') of an event A, which is simply the function that is 0 on A and 1 on A', can be expressed as $(f_1 - f_A) = (1 - f_A)$ the universal proposition minus f_A. The disjunction (∨) of two propositions f_A and f_B defined respectively on domains A and B is the indicator function $f_{A \vee B}$ defined by

$$f_{A \vee B}(\omega) = \begin{cases} 1, & \text{if } \omega \in A \cup B, \\ 0, & \text{if } \omega \notin A \cup B \end{cases} \quad (4.6)$$

This disjunction (∨) of two propositions A and B can be expressed as the maximum $\max(f_A, f_B)$ of the two indicator function f_A, and f_B. Similarly conjunction (∧) is $\min(f_A, f_B)$. For notational simplicity, a proposition f_A will be denoted simply as "A" but will retain the indicator function meaning.

The probability $P(f_A)$ of a proposition or event f_A is defined to be $P(A)$, the probability of the P-measurable event A on which f takes the value 1. Therefore $P(f_A) = P(f_A = 1) = P(f_A^{-1}(1)) = P(\{\omega \in \Omega : f_A(\omega) = 1\})$.

4.5 Conditional Propositions, Events and Restricted Indicator Functions

Following De Finetti [90, 59] a conditional (A|B), "A given B" or "A if B", is an ordered pair of propositions or events with three possible truth states: (A|B) takes the truth value of A when B is true but (A|B) is "undefined" or "inapplicable" when B is false. That is,

$$(A \mid B) \text{ is } \begin{cases} \text{true} & \text{if A and B are true,} \\ \text{false} & \text{if A is false and B is true,} \\ \text{Undefined} & \text{if B is false.} \end{cases} \quad (4.7)$$

While De Finetti's 3-valuedness for conditionals is followed, the interpretation here of the third truth value as "undefined" or "inapplicable" differs markedly from that of De Finetti, who interprets the third truth value as "unknown" and so therefore as something similar or equivalent to a probability value between 0 (false) and 1 (true). By contrast the "inapplicable" interpretation is not a truth or falsity value;

it is an indicator of irrelevance. This crucial difference in interpretation leads to a difference in operations.

A conditional can be represented as a restricted (partially defined) indicator function, (A|B):

$$(A|B)(\omega) = \begin{cases} 1 & \text{if } \omega \in A \cap B, \\ 0 & \text{if } \omega \in A' \cap B, \\ \text{Undefined} & \text{if } \omega \in B' \end{cases}$$

$$= \begin{cases} A(\omega) & \text{if } \omega \in B, \\ \text{Undefined}, & \text{if } \omega \notin B \end{cases} \quad (4.8)$$

Since $B(\omega) = 1$ if $\omega \in B$, the latter can be expressed as

$$(A|B)(\omega) = \begin{cases} A(\omega) \wedge B(\omega), & \text{if } \omega \in B, \\ \text{Undefined}, & \text{if } \omega \notin B \end{cases} \quad (4.9)$$

Thus (A|B) is just the indicator function (A ∧ B) restricted to the instances $\omega \in B$.

For any conditional (A|B) with $P(B) \neq 0$, the conditional probability P(A|B) is defined as usual to be $P(A \wedge B) / P(B)$. With this definition, the conditionals (A|B) have conditional probabilities that also satisfy the 6 qualitative axioms of Suppes and Zanotti [91 or 92] for a conditional probability measure.

4.6 Extended Operations on Conditional Propositions

We can expand the definition of a conditional to include cases in which A and B themselves are conditionals. To do this we need only decide on the definition of a conditional whose premise is undefined (U), the other cases being already determined. We will interpret an undefined condition to mean that there is no additional restriction imposed by it:

$$[(A|B) | (C|D)](\omega) = [(A|B)(\omega) | (C|D)(\omega)]$$

$$= \begin{cases} (A|B)(\omega) & \text{if } (C|D)(\omega) \neq 0, \\ \text{Undefined} & \text{if } (C|D)(\omega) = 0 \end{cases}$$

$$= \begin{cases} (A|B)(\omega) & \text{if } \omega \in C \vee D', \\ \text{Undefined}, & \text{if } \omega \notin C \vee D' \end{cases}$$

$$= \begin{cases} A & \text{if } \omega \in B \wedge (C \vee D'), \\ \text{Undefined}, & \text{if } \omega \notin B \wedge (C \vee D') \end{cases} \quad (4.10)$$

So

$$(A|B) | (C|D) = (A | B(C \vee D')) \quad (4.11)$$

With the definition of a conditional event, and using the extended definitions of the operations on functions, definitions can be developed for disjunction (∨), conjunction (∧) and negation (′) to go along with division (|) as follows. Also see Calabrese [65].

$$[(A|B) \vee (C|D)](\omega) = (A|B)(\omega) \vee (C|D)(\omega)$$

$$= \begin{cases} (A(\omega) \wedge B(\omega)) \vee (C(\omega) \wedge D(\omega)) & \text{if } x \in B \cup D, \\ \text{Undefined} & \text{if } x \notin B \cup D \end{cases} \quad (4.12)$$

The latter expression is just the conditional $((A \wedge B) \vee (C \wedge D) | (B \vee D))$. Thus

$$(A|B) \vee (C|D) = ((AB \vee CD) | (B \vee D)) \quad (4.13)$$

Here, juxtaposition of events A and B has replaced the conjunction notation $A \wedge B$. $(A|B) \vee (C|D)$ is just $(AB \vee CD)$ restricted to $(B \vee D)$.

For example, consider the experiment of rolling an ordinary 6-sided die once, and observing the number n showing up on the die. Suppose a wager is made that "if n is even then it will be a 2, or if n < 5 then n < 4". Each of the two component conditionals is applicable on a different subset of outcomes of the die roll, and combining them with "or" results in a disjunction of two conditional propositions.

By using 4.13) this disjunction is equivalent to a single conditional, with a conditional probability: (n=2 | n is even) ∨ (n < 4 | n < 5) = ((n=2) ∨ (n<4) | (n ≠ 5)) = ({1,2,3} | {1,2,3,4,6}), which is the conditional event that if the roll is not 5 then it will be 1, 2 or 3. This has conditional probability 3/5. By brute force examination of the 6 outcomes, this result can be seen to be consistent with intuition: Only a non-5 is applicable to at least one of the two component conditionals. So a "5" roll doesn't count. Given a non-5 roll the set {1,2,3} corresponds to winning the wager since "1" and "3" satisfy the second component while "2" satisfies the first component, but "4" and "6" satisfy neither component.

Similarly, for conjunction ($\wedge$)

$[(A|B) \wedge (C|D)](\omega) = (A|B)(\omega) \wedge (C|D)(\omega)$

$$= \begin{cases} (ABCD)(\omega) \text{ if } \omega \in B \cap D, \\ (AB)(\omega) \text{ if } \omega \in B \cap D', \\ (CD)(\omega) \text{ if } \omega \in B' \cap D, \\ \text{Undefined if } \omega \in B' \cap D' \end{cases} = \begin{cases} (ABCD)(\omega) \text{ if } \omega \in B \cap D, \\ (ABD')(\omega) \text{ if } \omega \in B \cap D', \\ (B'CD)(\omega) \text{ if } \omega \in B' \cap D, \\ \text{Undefined if } \omega \in B' \cap D' \end{cases}$$

$$= \begin{cases} (ABCD \vee ABD' \vee B'CD)(\omega) & \text{if } \omega \in B \cup D, \\ \text{Undefined} & \text{if } \omega \notin B \cup D \end{cases} \quad (4.14)$$

Thus
$$(A|B) \wedge (C|D) = (ABCD \vee ABD' \vee B'CD) \mid (B \vee D) \quad (4.15)$$

The negation operation is $[(A|B)'](\omega) = [(A|B)(\omega)]' = A'(\omega)$ if $\omega \in B$, or undefined if $\omega \in B'$. So

$$(A|B)' = (A'|B) \quad (4.16)$$

This algebra of uncertain conditional events or propositions has been extensively developed in [63-69] including a theory of deduction for uncertain conditionals extending Boolean deduction. See [63, p. 227] for an account of the Boolean properties retained and lost in the algebra of conditionals.

Concerning the structure of this algebra of conditionals, the conjunction and disjunction operations are obviously commutative and idempotent. They are less obviously also associative. The inapplicable conditional (1|0) is the unique absolute unit since for all conditionals $(x|y)$, $(x|y) \wedge (1|0) = (x|y)$ and $(x|y) \vee (1|0) = (x|y)$. While there is a unique *relative* complement $(a'|b)$ for each conditional $(a|b)$ such that $(a|b) \vee (a'|b) = (1|b)$ and $(a|b) \wedge (a'|b) = (0|b)$, there are no absolute complements. Although $(x|y) \wedge 0 = 0$ and $(x|y) \vee 1 = 1$, it is also true that $(x|y) \vee 0 = xy$ and $(x|y) \wedge 1 = x \vee y'$. Neither distributive law holds in general. However, conjunction distributes over disjunction if and only if whenever the outside conditional is true and one of the inside conditionals is false, then the other inside conditional is applicable. Similarly, disjunction distributes over conjunction if and only if whenever the outside conditional is false and one of the inside conditionals is true then the other inside conditional is applicable. (A proof of these facts about distributivity will be provided in a Section 8.2.4-5)

4.7 Weighted Averages

Among the interesting properties of these operations are the following weighted average formulas (See [67] p.1682) for the probabilities of the compound conditionals of equations 4.13 and 4.15:

$$P((A|B) \vee (C|D)) = P(B| B \vee D) P(A|B) + P(D| B \vee D) P(C|D)$$
$$- P(ABCD| B \vee D) \qquad (4.17)$$

$$P((A|B) \wedge (C|D)) = P(B| B \vee D) P(AD'|B) + P(D| B \vee D) P(CB'|D)$$
$$+ P(ABCD| B \vee D) \qquad (4.18)$$

The last term of Equations 4.17 and 4.18 can be written as $P(BD | B \vee D) P(AC | BD)$.

If the truth of each conditional implies the inapplicability of the other conditional, that is if AB is a subset of D' and CD is a subset of B', as for example when the two conditions, B and D, are disjoint, then both Equations 4.17 and 4.18 reduce to:

$$P((A|B) \vee (C|D)) = P((A|B) \wedge (C|D))$$
$$= P(B| B \vee D) P(A|B) + P(D| B \vee D) P(C|D) \qquad (4.19)$$

In any case, without any extra assumptions the following logical equation always holds:

$$(A|B) \vee (C|D) = (B| B \vee D) (A|B) \vee (D| B \vee D) (C|D) \qquad (4.20)$$

The right hand side of Equation 4.19 is a weighted average of the conditional probabilities of (A|B) and (C|D) where the weights are the probabilities of B and of D given either occurs. Because the conditional expectation of a restricted indicator function of an event A equals the conditional probability of A given the restriction B, this formula is equivalent to the one displayed by Suppes and Zanotti [91, p.165 or 92, p. 13] for the expectation of the disjunction of restricted indicator functions.

Note that because it is a weighted average the right hand side of Equation 4.19 will in general lie between P(A|B) and P(C|D) not above both. So disjunction of conditional events is not always monotonic; P((A|B) v (C|D)) can be less than P(A|B). Similarly P((A|B) ∧ (C|D)) can be greater than P(A|B). This is not strange because in general disjunction or conjunction of a conditional (A|B) with another conditional (C|D) expands the context to (B v D), which allows for greater or lesser probability than before the application of the operation: If (C|D) is (0|Ω) then disjunction with (A|B) yields AB, whose generally lower probability than P(A|B) is P(A|B) P(B). If (C|D) is (1|Ω) then disjunction with (A|B) yields (1| Ω), with probability 1.

As a simple example of this non-monotonicity, let Ω = {1,2,3,4,5,6}, the numbered faces of a 6-sided die thrown once, and let B = {2,4,6} and A = {2,4}. The conditional probability of rolling 2 or 4, given the roll is an even number, equals 2/3. That is, P(A | B) = 2/3. Now suppose also that C = {1} and D = {1,3,5}. So P(C|D) = 1/3. That is the probability of rolling a 1, given the roll is an odd number, equals 1/3. Now what is the probability of "rolling a 2 or 4 given the roll is even, OR rolling 1 given the roll is odd"? That is, P((A|B) v (C|D)) = ? The answer is P(AB v CD | B v D) = P{1,2,4} / P(even or odd) = 3/6 =1/2. So here P((A|B) v (C|D)) is less than P(A | B) alone.

4.8 Extended Operations on Random Variables

Having extended the operations for functions with different domains and having applied them to extend the operations for conditional propositions, it is possible to extend the operations on random variables and conditional random variables. A real-valued random variable X is a function from a sample space Ω of a probability space (Ω, $\mathcal{B}$, P) into the real numbers such that for any real number x, the set of instances ω ∈ Ω for which X(ω) < x is a member of $\mathcal{B}$, and so has a probability P{ω ∈ Ω: X(ω) < x }. It follows that there is a probability that X takes a value in any of the collection of Borel subsets of real numbers, consisting of those subsets that are a countable collection of intersections or unions of the intervals (-∞, x) or their complements [x,

∞), for any real number x. Of course any interval (x, y) of real numbers is a Borel set.

As with functions, just doing division on random variables in general produces new ones with different domains whenever the divisor assumes the real value 0. Subsequently, using standard techniques, operating with this restricted variable will propagate its restricted domain. However using these extended operations, the domains of functions can be expanded as well as restricted.

While the ordered pair (A|B) for events A, B, is defined and interpreted as "event A given event B is true", the corresponding construction (X|Y), where X and Y are random variables, can not be immediately interpreted because "given Y is true" does not make sense for real-valued random variables. The condition must be an event such as $Y \in B$, the event that Y takes a value in a Borel set of real numbers B.

4.9 Conditional Random Variables

Let X, Y, W, Z be real-valued random variables on a probability space $\mathcal{P} = (\Omega, \mathcal{B}, P)$ and let A, B, C, D, ... be Borel sets on the real line. A conditional random variable $(X \mid Y \in B)$ is just the random variable X restricted to the instances ω for which $Y(\omega) \in B$. That is,

$$(X \mid Y \in B) = \begin{cases} X(\omega), & \text{if } Y(\omega) \in B, \\ \text{Undefined}, & \text{if } Y(\omega) \notin B \end{cases} = X \text{ on } Y^{-1}(B) \quad (4.21)$$

If $Y^{-1}(B)$ is empty, then $(X \mid Y \in B)$ is completely undefined, defined for no instances ω.

Although conditional probability distributions, conditional density functions and conditional expectations have standard definitions (See, for instance Papoulis [93]), the operations of summation, difference, multiplication and division on conditional random variables are all expressed in terms of probability distributions rather than directly. But now these can follow directly from the extended definitions for operations on functions.

4.10 Operations on Conditional Random Variables

Using the extended definitions for operations on real-valued functions, extended operations for random variables can be defined as follows:

$[(X \mid Y \in B) + (W \mid Z \in D)](\omega)$

$$= \begin{cases} X(\omega) + W(\omega), & \text{if } Y(\omega) \in B \text{ or } Z(\omega) \in D, \\ \text{Undefined}, & \text{if } Y(\omega) \notin B \text{ and } Z(\omega) \notin D \end{cases} \quad (4.22)$$

Replacing "+" in 18) with negation (-), or multiplication (*) yields the corresponding operations on the two conditional random variables. Division requires a separate formula due to possible division by zero:

$[(X \mid Y \in B) \div (W \mid Z \in D)](\omega)$

$$= \begin{cases} X(\omega) \div W(\omega), & \text{if } Y(\omega) \in B \text{ or } Z(\omega) \in D, \\ \text{Undefined}, & \text{if } Y(\omega) \notin B \text{ and } Z(\omega) \notin D \end{cases}$$

$$= \begin{cases} X(\omega)/W(\omega) & \text{if } Y(\omega) \in B \text{ and } 0 \neq Z(\omega) \in D, \\ X(\omega) & \text{if } Y(\omega) \in B \text{ and } Z(\omega) \notin D, \\ 1/W(\omega) & \text{if } Y(\omega) \notin B \text{ and } 0 \neq Z(\omega) \in D, \\ \text{Undefined} & \text{if } Y(\omega) \notin B \text{ and } (Z(\omega) = 0 \text{ or } Z(\omega) \notin D) \end{cases} \quad (4.23)$$

4.11 Expectations and Conditional Expectations

The expectation or average E(X) of a random variable X is defined to be just the sum of the values of X each weighted by its probability. Keeping to an elementary formulation for simplicity of exposition, assume the universe Ω is finite or countable. Then

$$E(X) = \sum_{\omega \in \Omega} X(\omega) P(\omega) \quad (4.24)$$

If X and W are two random variables defined on Ω then easily E(X + W) = E(X) + E(W).

By standard definitions, the conditional expectation $E(X \mid Y \in B)$, where Y is a random variable on Ω and B is a Borel subset of real numbers, is defined to be

$$E(X \mid Y \in B) = E(X \text{ on } Y^{-1}(B)) = \sum_{\omega \in \Omega} X(\omega) P(\omega \mid Y^{-1}(B))$$

$$= \sum_{\omega \in \Omega} X(\omega) \, P(\omega \wedge Y^{-1}(B)) / P(Y^{-1}(B))$$

$$= [1 / P(Y \in B)] \sum_{Y(\omega) \in B} X(\omega) \, P(\omega) \qquad (4.25)$$

Note here that if $P(Y^{-1}(B)) = 0$, then the conditional expectation is undefined. Otherwise, $P(\omega \mid Y^{-1}(B)) = 0$ for $\omega \notin Y^{-1}(B)$ and $P(\omega \mid Y^{-1}(B)) = P(\omega) / P(Y^{-1}(B))$ for $\omega \in Y^{-1}(B)$.

$E(X \mid Y \in B)$ is just the expectation of the random variable X restricted to the instances ω for which $Y(\omega)$ takes a value in B. The individual probabilities of these instances is just normalized by $P(Y \in B)$ so that their sum is 1 while they maintain the same relative probabilities with respect to each other as before the conditioning.

Now it is well known (see for example Calabrese [93], p.144) that if $Y^{-1}(B)$ and $Z^{-1}(D)$ are disjoint and exhaustive of Ω, i.e, $Y^{-1}(B) \wedge Z^{-1}(D) = \Phi$ and $Y^{-1}(B) \vee Z^{-1}(D) = \Omega$ and if X, Y, W and Z are random variables on Ω, then

$$E(\,(X \mid Y \in B) + (W \mid Z \in D)\,)$$
$$= E(X \mid Y \in B)\, P(Y \in B) + E(W \mid Z \in D)\, P(Z \in B) \qquad (4.26)$$

That is, the expectation of the sum of the conditional random variables is the sum of the conditional expectations weighted by the probabilities of the associated conditions. With the extended definitions of operations on random variables this result can be generalized to allow $Y^{-1}(B)$ and $Z^{-1}(D)$ to be arbitrary events that may overlap and also may not be exhaustive of Ω.

First we extend the result to disjoint events $Y^{-1}(B)$ and $Z^{-1}(D)$ that do not necessarily exhaust Ω.

4.12 Lemma for Theorem 4.13

If $Y^{-1}(B)$ and $Z^{-1}(D)$ are disjoint events of Ω, and if X, Y, W and Z are random variables on Ω, then

$$E(\,(X \mid Y \in B) + (W \mid Z \in D)\,)$$
$$= E(X \mid Y \in B)\, P(Y \in B \mid Y \in B \vee Z \in D)$$
$$+ E(W \mid Z \in D)\, P(Z \in B \mid Y \in B \vee Z \in D) \qquad (4.27)$$

Proof of Lemma for Theorem 4.13 This result follows by using a new probability measure on just the part of Ω inside $(Y \in B \vee Z \in D)$. So let Q be the probability measure defined by $Q(A) = P(A \mid Y \in B \vee Z \in D)$ for any event A in Ω. That Q is a probability measure on $(Y \in B \vee Z \in D)$ is easy to show since it is non-negative, $Q(Y \in B \vee Z \in D) = P(Y \in B \vee Z \in D \mid Y \in B \vee Z \in D) = 1$, and finally, if A and C are disjoint events in Ω, then $Q(A \vee C) = P(A \vee C \mid Y \in B \vee Z \in D) = P(A \vee C) / P(Y \in B \vee Z \in D) = (P(A) + P(C)) / P(Y \in B \vee Z \in D) = Q(A) + Q(B)$. In addition, the conditional expectation $E_Q(X \mid Y \in B)$ with respect to Q of an arbitrary random variable X given arbitrary $(Y \in B)$ equals the conditional expectation $E_P(X \mid Y \in B)$ with respect to P because

$$\begin{aligned} E_Q(X \mid Y \in B) &= \sum_{\omega \in \Omega} X(\omega) \, Q(\omega \wedge (Y \in B) \mid Y \in B)) \\ &= \sum_{\omega \in \Omega} X(\omega) \, Q(\omega \wedge (Y \in B)) / Q(Y \in B) \\ &= \sum_{\omega \in \Omega} X(\omega) \, P(\omega \wedge (Y \in B)) / P(Y \in B)) \\ &= E_P(X \mid Y \in B). \end{aligned} \quad (4.28)$$

So now computing $E_P((X \mid Y \in B) + (W \mid Z \in D)) = E_Q((X \mid Y \in B) + (W \mid Z \in D)) =$

$$\begin{aligned} &E((X \mid Y \in B) \, Q(Y \in B) + E(W \mid Z \in D) \, Q(Z \in D) \\ &= E(X \mid Y \in B) \, P(Y \in B \mid Y \in B \vee Z \in D) \\ &\qquad + E(W \mid Z \in D) \, P(Z \in D \mid Y \in B \vee Z \in D) \end{aligned} \quad (4.29)$$

That completes the proof of the Lemma for Theorem 4.13.

4.13 Theorem (Expected Value of a Sum of Conditional Random Variables)
If X, Y, W and Z are real-valued random variables and B and D are arbitrary Borel subsets of real numbers, then

$$\begin{aligned} &E((X \mid Y \in B) + (W \mid Z \in D)) = \\ &E(X \mid Y \in B) \, P(Y \in B \mid Y \in B \vee Z \in D) \\ &\qquad + E(W \mid Z \in D)) \, P(Z \in D \mid Y \in B \vee Z \in D) \end{aligned} \quad (4.30)$$

Proof of Theorem 4.13 Let $K = Y^{-1}(B) = \{\omega \in \Omega : Y(\omega) \in B\}$ and $L = Z^{-1}(D)$. So using the definition of extended summation for conditional random variables,

$$E((X \mid Y{\in}B) + (W \mid Z{\in}D))$$
$$= E((X \mid K) + (W \mid L)) = E(X + W \mid K \vee L)$$
$$= E((X \mid KL') + (X + W \mid KL) + (W \mid K'L)) \quad (4.31)$$

where juxtaposition has again replaced conjunction ($\wedge$) to shorten notation. Since KL', KL and $K'L$ are disjoint, according to the Lemma for Theorem 4.13, we can continue with

$$E((X \mid KL') + (X + W \mid KL) + (W \mid K'L))$$
$$= E(X \mid KL')P(KL' \mid K \vee L) + E(X + W \mid KL)\,P(KL \mid K \vee L)$$
$$+ E(W \mid K'L)\,P(K'L \mid K \vee L) \quad (4.32)$$

$$= E(X \mid KL')P(KL' \mid K \vee L) + E(X \mid KL)\,P(KL \mid K \vee L)$$
$$+ E(W \mid KL)\,P(KL \mid K \vee L)) + E(W \mid K'L)\,P(K'L \mid K \vee L) \quad (4.33)$$

$$= E((X \mid KL')\,P(KL' \mid K)\,P(K \mid K \vee L)$$
$$+ E(X \mid KL)\,P(KL \mid K)\,P(K \mid K \vee L)$$
$$+ E(W \mid KL)\,P(KL \mid L)\,P(L \mid K \vee L)$$
$$+ E(W \mid K'L)\,P(K'L \mid L)\,P(L \mid K \vee L) \quad (4.34)$$

$$= [E((X \mid KL')P(KL' \mid K) + E(X \mid KL)\,P(KL \mid K)]\,P(K \mid K \vee L)$$
$$+ [E(W \mid KL)\,P(KL \mid L) + E(W \mid K'L)\,P(K'L \mid L)]\,P(L \mid K \vee L) \quad (4.35)$$

$$= E((X \mid KL') + (X \mid KL))\,P(K \mid K \vee L)$$
$$+ E((W \mid KL) + (W \mid K'L))\,P(L \mid K \vee L) \quad (4.36)$$

using the Lemma for Theorem 4.13 in reverse. Thus

$$E((X \mid K) + (W \mid L))$$
$$= E(X \mid K)\,P(K \mid K \vee L) + E(W \mid L)\,P(L \mid K \vee L) \quad (4.37)$$

That is,
$$E((X \mid Y{\in}B) + (W \mid Z{\in}D))$$
$$= E(X \mid Y{\in}B)\,P(Y{\in}B \mid Y{\in}B \vee Z{\in}D)$$
$$+ E(W \mid Z{\in}D))\,P(Z{\in}D \mid Y{\in}B \vee Z{\in}D) \quad (4.38)$$

That completes the proof of Theorem 4.13.

4.14 Theorem (Expected Value of a Product of Conditional Random Variables)

If X, Y, W and Z are real-valued random variables and B and D are arbitrary Borel subsets of real numbers, then the expectation of the product of the conditional random variables (X | Y∈B) and (W | Z∈D) is given by

$$E((X | Y \in B) * (W | Z \in D)) =$$
$$E(X | Y \in B \land Z \notin D) \, P(Y \in B \land Z \notin D | Y \in B \lor Z \in D)$$
$$+ \, E(X * W | Y \in B \land Z \in D) \, P(Y \in B \land Z \in D | Y \in B \lor Z \in D)$$
$$+ \, E(Z | Y \notin B \land Z \in D) \, P(Y \notin B \land Z \in D | Y \in B \lor Z \in D) \quad (4.39)$$

Proof of Theorem 4.14. By the extended definition of products,

$$(X | Y \in B) * (W | Z \in D) = \begin{cases} X & \text{if } Y \in B \text{ and } Z \notin D, \\ X*W & \text{if } Y \in B \text{ and } Z \in D, \\ W & \text{if } Y \notin B \text{ and } Z \in D, \\ \text{Undefined} & \text{if } Y \notin B \text{ and } Z \notin D \end{cases} \quad (4.40)$$

where the domain of the product random variable has been broken into disjoint events. Then by the definition of the conditional expectation (Equation 4.25), the expectation of this product random variable, E((X | Y∈B) * (W | Z∈D)), is immediately expressed by Equation 4.39. This completes the proof of Theorem 4.14.

With a kind of independence, Equation 4.39 can be somewhat simplified. Recall that two random variables, X and Z, are independent if P(X∈A and Z∈C) = P(X∈A) P(Z∈C) for any events X∈A and Z∈C. Knowing the value taken by one variable does not change the probability of the other variable taking its values.

4.15 Definition of Independence of Random Variables

Two random variables X and W are independent if they are independent on each common domain. That is, X and W are independent if for any event H for which both X and W are defined, X is conditionally independent of W given H. That is

$$P(X \in A \land W \in C | H) = P(X \in A | H) \, P(W \in C | H).$$

4.16 Corollary to Theorem 4.14 for Independent Variables
If X and W are independent random variables then under the hypothesis of Theorem 4.14,

$$E((X \mid Y \in B) * (W \mid Z \in D))$$
$$= E(X \mid Y \in B \wedge Z \not\in D) \, P(Y \in B \wedge Z \not\in D \mid Y \in B \vee Z \in D)$$
$$+ E(X \mid Y \in B \wedge Z \in D) \, E(W \mid Y \in B \wedge Z \in D) \, P(Y \in B \wedge Z \in D \mid Y \in B \vee Z \in D)$$
$$+ E(W \mid Y \not\in B \wedge Z \in D) \, P(Y \not\in B \wedge Z \in D \mid Y \in B \vee Z \in D) \quad (4.41)$$

Proof of Corollary to Theorem 4.14 It is well known that the expectation of a product of independent random variables is the product of the expectations. Therefore $E(X * W \mid Y \in B \wedge Z \in D) = E(X \mid Y \in B \wedge Z \in D) \, E(W \mid Y \in B \wedge Z \in D)$, and the result follows by substitution into Equation 4.39.

4.17 Work Force Example
Consider a work force consisting of workers $i = 1, 2 \ldots n$ with variable work output levels $W_1, W_2 \ldots W_n$ and work shifts $s_1, s_2 \ldots s_n$ respectively spanning the 24 hour day. To formulate the problem in terms of random variables, let $s_n(\omega) = 1$ if time $\omega \in s_i$ and 0 otherwise. Then the work level at time ω of worker i is $(W_i \mid s_i(\omega) = 1)$. The sum of work output of all workers is $\sum_i (W_i \mid s_i(\omega) = 1)$, and the average or expected work level over the day is $E(\sum_i (W_i \mid s_i(\omega) = 1)) = \sum_i E(W_i \mid s_i(\omega) = 1) \, P(s_i(\omega) = 1)$.

4.18 Surveillance Region Example
Let B and C be the surveillance regions of two radars, R1 and R2, and suppose $X(\omega)$ is the error rate of missed detections by R1 at any place $\omega \in B$, and $W(\omega)$ is the error rate by R2 at any place $\omega \in C$. X and W are undefined outside their respective domains B and C. Then using the definition of extended product, and assuming independence of detections by R1 and R2, $(X * W)(\omega) = X(\omega) W(\omega)$ is the combined error rate of missed detections by both radars over $(B \cup C)$. This combined error rate is X on $B \cap C'$, $X * W$ on $B \cap C$, and W on $B' \cap C$.

Now suppose in addition that the detection rate of radar R1 is greatly affected by fog F while interrogation radar R2 is most affected by the density D of communication on interrogation frequencies.

Measuring fog as "heavy (h), medium (m), or none (n)" and communication traffic density on a scale from 1 to 3, the error rate over (B ∪ C) under conditions of heavy fog and communication density 2 is $((X * W) \mid (F=h) \wedge (D=2) \wedge (B \cup C)) = ((X * W) \mid (F=h)(D=2)(B \cup C))$. So the expected combined error rate of the two radars given heavy fog and medium (2) communication density is $E((X * W) \mid (F=h)(D=2)(B \cup C))$.

Now by the product definition

$$((X * W) \mid (F=h)(D=2)(B \cup C)) = \\ (X \mid (F=h)(D=2) BC') \\ \vee (X * W \mid (F=h)(D=2) BC) \vee (W \mid (F=h)(D=2) B'C) \quad (4.42)$$

Since the detection errors for the two radars are assumed independent, the last equation simplifies to

$$((X * W) \mid (F=h)(D=2)(B \cup C)) = \\ (X \mid (F=h) BC') \\ \vee (X * W \mid (F=h)(D=2) BC) \vee (W \mid (D=2)B'C) \quad (4.43)$$

Let $G = (F=h)(D=2)(B \cup C)$. Then in terms of the average error rates of the individual radars, the average combined error rate given heavy fog and medium (2) communication density over the combined surveillance region B ∪ C) is

$$E((X * W) \mid (F=h)(D=2)(B \cup C)) \\ = E(X \mid (F=h)BC') P((F=h)BC') \mid G) \\ + E(X*W \mid (F=h)(D=2) BC) P((F=h)(D=2)BC \mid G) \\ + E(W \mid (D=2)(BC')) P(D=2) B'C \mid G) \quad (4.44)$$

$$= E(X \mid (F=h)(BC')) P((F=h)(BC') \mid G) \\ + E(X \mid (F=h)(BC)) E(W \mid (D=2)(BC)) P((F=h)(D=2)(BC) \mid G) \\ + E(W \mid (D=2)(B'C)) P((D=2)(B'C) \mid G) \quad (4.45)$$

using conditional independence again to split the expectation of X*W and to simplify the conditions.

For simplicity, assume that B ∪ C is the whole universe and that fog is heavy (F=h) everywhere and communication density is medium

(D=2) everywhere in $B \cup C$. So G = the whole universe Ω, and $P((F=h)BC') | G) = P(BC')$. Similarly, $P((F=h)(D=2)BC | G) = P(BC)$ and $P((D=2)(B'C) | G) = P(B'C)$. Thus

$$E((X * W) | (F=h) (D=2) (B \cup C))$$
$$= E(X | (F=h)(BC')) P(BC')$$
$$+ E(X | (F=h)(BC)) E(W | (D=2)(BC)) P(BC)$$
$$+ E(W | (D=2)(B'C)) P(B'C) \quad (4.46)$$

If the error rate of radar R1 is 0.04 in heavy fog (F=h) and the error rate of R2 is 0.02 in medium communication density (D=2), the combined error rate under the conditions is

$$E((X * W) | (F=h) (D=2) (B \cup C))$$
$$= (0.04) P(BC') + (0.04)(0.02) P(BC) + (0.02) P(B'C) \quad (4.47)$$

Note that the error rates are multiplied in the common surveillance region BC where the combined error rate is just 0.0008.

4.19 Summary

Extended definitions of function addition and other operations have been applied to conditional propositions and conditional events, and to conditional random variables. This allows direct manipulation of conditional events and of conditional random variables without resort to a probability or density function. General formulas for the expectation of the sum, and of the product, of two conditional random variables have been determined. Finally two examples illustrate the use of these formulas in practical situations.

Chapter 5
The Structure of Conditional Logic Probability

> The set S of all sentences *is the intersection of all those sets* which contain *all sentential variables* (elementary sentences) and are closed under the operations of forming implications and negations.
>
> J. Lukasiewicz & A. Tarski [181],
> Investigations into the Sentential Calculus, 1933

5.1 Ideals of Propositions

In algebraic logic a condition or set of conditions refers to an ideal of assumed (always) true propositions. The set of all always-true propositions, that is, the set of theorems, is an ideal in the following sense.

5.1.1 Definition of Ideals (filters, sum ideals) of Propositions

A subset I of propositions is an ideal (filter, sum ideal) of propositions if and only if the following two statements hold,

$$\text{If } p \in I \text{ and } q \in I \text{ then } (p \wedge q) \in I \qquad (5.1.a)$$
$$\text{If } p \in I \text{ and } q \text{ is any proposition in L then } (p \vee q) \in I \qquad (5.1.b)$$

Equivalently, I is an ideal if

$$(p \wedge q) \in I \text{ if and only if } p \in I \text{ and } q \in I \qquad (5.1.c)$$

The ideal L of all propositions is said to be improper because not every proposition should be always true. The logic should be consistent.

Let A be the implicitly or explicitly assumed axioms of L, and let T be the (not necessarily maximal) ideal of (always) true propositions generated by A. That is, T is the smallest ideal that includes A. Any ideal I induces an equivalence relation $\equiv$ on L as follows:

5.1.2 Definition of Equivalent Propositions Given an Ideal
Two propositions, p and q, are equivalent given an ideal I if and only if there is some proposition r in I such that $p \wedge r = q \wedge r$. In symbols,

$$(p|I) \equiv (q|I) \text{ if and only if } p \wedge r = q \wedge r \text{ for some } r \in I. \quad (5.1.d)$$

Since there will be no confusion between "=", the initial Boolean equivalence relation, and " ≡ ", the equivalence arising from the conditional adjunction of an ideal I, it is convenient to denote both equivalences by " = ".

5.1.3 Theorem: Relation of Definition 5.1.2 is an Equivalence Relation.
Proof of Theorem 5.1.3 Reflexivity follows because I is nonempty. (All ideals contain 1.) Symmetry follows by the commutative law of conjunction. Transitivity follows because if $(p|I) = (q|I)$ and $(q|I) = (r|I)$, then there are propositions $s \in I$ and $t \in I$ such that $p \wedge s = q \wedge s$ and $q \wedge t = r \wedge t$. But then $p \wedge (s \wedge t) = (p \wedge s) \wedge t = (q \wedge s) \wedge t = (q \wedge t) \wedge s = (r \wedge t) \wedge s = r \wedge (s \wedge t)$, where $(s \wedge t)$ is in I. That is, $(p|I) = (r|I)$. This completes the proof of Theorem 5.1.3.

The set of all equivalence classes of propositions of L generated by an ideal I will be denoted L/I. Let (p|I) denote the equivalence class containing p. One advantage of Definition 5.1.2 is the explicit use of the ideal I of always true propositions, relative to which two propositions p and q may be equivalent.

5.2 The Conditional Logic Generated by an Ideal
L/I is the conditional logic resulting from the additional assumption of the propositions of I.

Theorem 5.2.1 Equivalence Classes Preserved
The equivalence classes given an ideal I are preserved under the operations of $\wedge$, $\vee$, and $\sim$. Thus L/I is a homomorphic image of L by the natural mapping

$$p \rightarrow (p|I), \quad (5.2.a)$$

$$(p|I) \wedge (q|I) = (p \wedge q \mid I), \tag{5.2.b}$$

$$(p|I) \vee (q|I) = (p \vee q \mid I), \tag{5.2.c}$$

$$\sim(p \mid I) = (\sim p \mid I). \tag{5.2.d}$$

Proof of Theorem 5.2.1 Suppose $(p|I) = (p_1|I)$ and $(q|I) = (q_1|I)$. So there are two propositions r and s in I such that $p \wedge r = p_1 \wedge r$ and $q \wedge s = q_1 \wedge s$. It then easily follows, using that $(r \wedge s)$ is in I, that $(p \wedge q \mid I) = (p_1 \wedge q_1 \mid I)$. Similarly, given I, $p \vee q$ is equivalent to $p_1 \vee q_1$. Finally, if $(p|I) = (q|I)$ and r is in I with $p \wedge r = q \wedge r$, then

$$\begin{aligned}
r \wedge \sim p &= (r \wedge (q \vee \sim q)) \wedge \sim p \\
&= [(r \wedge q) \vee (r \wedge \sim q)] \wedge \sim p \\
&= (r \wedge q \wedge \sim p) \vee (r \wedge \sim p \wedge \sim q) \\
&= (r \wedge p \wedge \sim p) \vee (r \wedge \sim p \wedge \sim q) \\
&= 0 \vee (r \wedge \sim p \wedge \sim q) \\
&= r \wedge \sim p \wedge \sim q
\end{aligned}$$

Therefore $(r \wedge \sim p) \leq \sim q$. And so $(r \wedge \sim p) \leq r \wedge \sim q$. By symmetry, $(r \wedge \sim q) \leq (r \wedge \sim p)$. Therefore $(r \wedge \sim q) = (r \wedge \sim p)$. So $\sim p = \sim q$ given I.

More directly, this can be proved as follows:

$$\begin{aligned}
\sim p \wedge r &= \sim p \wedge r \wedge (q \vee \sim q) \\
&= (\sim p \wedge r \wedge q) \vee (\sim p \wedge r \wedge \sim q) \\
&= (\sim p \wedge r \wedge p) \vee (\sim p \wedge r \wedge \sim q) \\
&= (q \wedge r \wedge \sim q) \vee (\sim p \wedge r \wedge \sim q) \\
&= (p \wedge r \wedge \sim q) \vee (\sim p \wedge r \wedge \sim q) \\
&= (\sim p \vee p) \wedge (r \wedge \sim q) \\
&= \sim q \wedge r.
\end{aligned}$$

This completes the proof of Theorem 5.2.1.

5.2.2 Definition of the Conditional Logic L/I, Given an ideal I

The conditional logic L/I, given an ideal I, is the Boolean logic of equivalence classes of propositions $\{(q|I): q \in L\}$ defined by Equations (5.1.d and (5.2.b-d).

If p is a proposition, then (p) denotes the smallest ideal that includes p. In L/(p) the equivalence class containing q is denoted (q|p), q given p. For I = (p) Theorem 5.2.1 yields

$$(q|p) \vee (s|p) = (q \vee s \mid p) \quad (5.2.e)$$
$$(q|p) \wedge (s|p) = (q \wedge s \mid p) \quad (5.2.f)$$
$$\sim (q|p) = (\sim q \mid p) \quad (5.2.g)$$

and Definition 5.1.2 yields

$$(q|p) = (q \wedge p \mid p). \quad (5.2.h)$$

(q|p) will be said to be in reduced form if $q \wedge p = q$.

5.2.3 Theorem: Logic L is isomorphic to L/(1)
L is isomorphic to L/(1) by the mapping

$$p \leftrightarrow (p \mid 1). \quad (5.2.i)$$

Proof of Theorem 5.2.3 (p|1) = (q|1) if and only if p = q, since $p = p \wedge 1 = q \wedge 1 = q$.

By the identification (5.2.i), the equivalence (p|I) = (q|I) reverts to p = q when the conditioning ideal I is the set of all previously assumed or provable propositions, namely (1).

5.3 Logical and Set-Theoretical Operations on Ideals
In algebraic logic a proposition p is represented by the ideal (p) that it generates. Operations on propositions become operations on ideals.

5.3.1 Definition of Logical Operations on Ideals
If I and J are two sets of propositions of L, then define

$$I \wedge J = \{p \wedge q : p \in I, q \in J\}, \quad (5.3.a)$$

$$I \vee J = \{p \vee q : p \in I, q \in J\}, \quad (5.3.b)$$

$$I \cap J = \{p : p \in I \text{ and } p \in J\}, \quad (5.3.c)$$

$$I \cup J = \{p: p \in I \text{ or } p \in J\}, \qquad (5.3.d)$$

$$\sim I = \{\sim p: p \in I\}, \qquad (5.3.e)$$

$$I' = \{p: p \notin I\}. \qquad (5.3.f)$$

5.3.2 Theorem: I ∩ J, I ∨ J, and I ∧ J are Ideals

If I and I are ideals, then $I \cap J$, $I \vee J$, and $I \wedge J$ are all ideals, but $I \cup J$ may not be an ideal. In any case

$$I \vee J = I \cap J \subset I \cup J \subset I \wedge J \qquad (5.3.g)$$

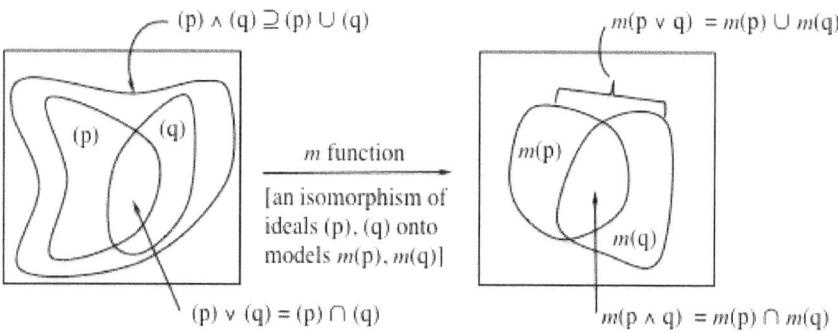

Figure 5.1 Disjunctions and Conjunctions of Propositional Ideals and Associated Models

The proof of Theorem 5.3.2 is straightforward by standard set-theory techniques and the properties (5.1.a-c) of ideals. (Also see [5, pp. 542-545].)

Although the disjunction of two propositional ideals is equivalent to the intersection of those ideals, the conjunction of two propositional ideals is often something more than the union of those ideals. The ideal whole is often greater than the union of its ideal parts.

5.3.3 Corollary Concerning ~I and I' for Ideals

In the context of Definition 5.3.1, if I is not the trivial ideal of all propositions L, then $\sim I \subseteq I'$ with equality if and only if I is maximal.

Proof of Corollary 5.3.3 By definition, $\sim I = \{\sim p: p \in I\}$. If $p \in I$, then $\sim p$ cannot also be in I since I≠L. It must be in I'. So $\sim I \subset I'$. Conversely, if I is maximal, then $\sim p \in I$ whenever $p \notin I$. But $\sim p \in I$ implies that $\sim(\sim p) \in \sim I$. That is, $p \in \sim I$. Thus $I' \subset \sim I$. Note that if $I=L$ then I' is empty but $\sim I = I$. That completes the proof.

By Theorem 5.3.2, $I \wedge J$ is obviously the ideal generated by two ideals I and J. Similarly, $I \cap J$ is clearly the largest ideal common to I and J, and $I \cap J = I \vee J$.

For propositions and ideals "and" includes "union". That is, $I \wedge J$ includes all the propositions of $I \cup J$. Somewhat in contrast, "or" corresponds exactly to "intersection". $I \vee J$ means those propositions that are common to both I and J. When probabilities are assigned via the extension function, these features of language will be explicitly incorporated into the general theory of probability.

Now that the operations have been extended to ideals, Definition 5.1.2 for equivalence can be extended to conditional ideals:

5.3.4 Definition: Equivalence of Ideals Given an Ideal I
If I, J, and K are ideals of L, then

$$(J|I) = (K|I) \text{ if and only if } J \wedge I = K \wedge I. \tag{5.3.i}$$

Thus

$$(J \wedge I \mid I) = (J \mid I). \tag{5.3.j}$$

$(J|I)$ will be said to be in reduced form if $J \wedge I = J$.

The relation of Definition 5.3.4 is easily seen to be an equivalence relation. More generally, equivalence (=) is defined as follows.

5.3.5 Definition: Equivalence of Conditional Ideals
For two arbitrary conditional ideals $(J|I)$ and $(K|H)$

$$(J|I) = (K|H) \text{ if and only if } I = H \text{ and } J \wedge I = K \wedge H$$

Note that if J, I, K, and H are principal ideals generated by propositions q, p, s, and r respectively, then

$$(q|p) = (s|r) \text{ if and only if } p = r \text{ and } q \wedge p = s \wedge r. \tag{5.3.k}$$

5.4 Models of a Boolean Propositional Logic
According to C. C. Chang and H. J. Keisler [14, p1-2], the models of $\mathcal{L}$ may be defined as follows:

5.4.1 Definition: The Models of a Logic $\mathcal{L}$

The models of $\mathcal{L}$ satisfy all the axioms of $\mathcal{L}$. In addition, if p is a proposition and ω is a model of $\mathcal{L}$, then either p is true in ω or else p is false in ω.

Thus each model ω defines a function from the set of all propositions L onto the set {true, false}.

According to Definition 5.4.1 each model is evidently the only model of some categorical axiom system whose axioms include the axioms of $\mathcal{L}$. In a model every proposition is either always true or always false.

5.5 Ideals of Models and Ideals of Conditional Events

Analogous to the ideals of L there are the set-theoretic ideals of Ω. The definition is like Definition 5.1.1:

5.5.1 Definition: Ideals of Models

A collection Δ of subsets A, B... of the universe of models, Ω, is a sum ideal if and only if

If $A \in \Delta$ and $B \in \Delta$ then $(A \cap B) \in \Delta$ (5.5.a)
If $A \in \Delta$ and B is any subset of Ω then $(A \cup B) \in \Delta$ (5.5.b)

In an entirely similar way conditional events (conditional models) arise. If Δ is a sum ideal of Ω, denote by Ω/Δ the system of equivalence classes determined by the equivalence relation

$(A \mid \Delta) = (B \mid \Delta)$ if & only if $A \cap C = B \cap C$ for some C in Δ   (5.5.c)

(A | Δ) denotes the equivalence class containing A. (A | B) denotes the equivalence class containing the subset A in the conditional collection of subsets, Ω/B, generated by the smallest ideal (B) containing the subset B. Thus if A, B, C, and D are subsets of Ω, then

$(A|C) = (B|D)$ if & only if $C = D$ and $A \cap C = B \cap D$ (5.5.d)

5.6 The Extension Mapping
The extension function m can now be more fully developed.

5.6.1 Theorem: The Extension Mapping Reverses Inclusion
If I and J are sets of propositions and $I \subset J$, then $m(I) \supset m(J)$. The propositions really do filter the models. See Figure 3.1.

Proof of Theorem 5.6.1 Any model that satisfies all the propositions of J satisfies all the propositions of I.

The following theorem justifies the use of the notation "m(I)" for both the extension of a set of propositions I and the extension of (I), the ideal generated by I.

5.6.2 Theorem: The Extension of a Set of Propositions and its Ideal
A set of propositions I, and the ideal (I) generated by I, have the same extension set of models.

Proof of Theorem 5.6.2 Since $I \subset (I)$, it follows from Theorem 5.6.1 that every model of (I) is a model of I. We need only show that any model in which all the propositions of I are true is also a model in which all the propositions of (I) are true. Let K= {all finite conjunctions of propositions in I}, and let J = {q ∨ p: q in L, p in K}. Then since (I) is the smallest ideal that contains I, it follows that J = (I), because J is an ideal and $J \subset (I)$. If ω is a model in which every proposition of I is true, then all finite conjunctions of statements of I are true in ω. And all disjunctions of such finite conjunctions are true in ω. That is, all the propositions of J are true in ω. Since J = (I), every model of I is a model of (I). This completes the proof of Theorem 5.6.2.

The following two theorems show that the extension function m is a homomorphism of L onto Ω. It is helpful to see this illustrated in the usual Venn diagram of a sample space of models. See Figure 5.2.

5.6.3 Theorem: The Extension of a Conjunction & Disjunction of Ideals
If I and J are sets of propositions of L, then

and
$$m(I \wedge J) = m(I) \cap m(J) \quad (5.6.a)$$
$$m(I \vee J) = m(I) \cup m(J). \quad (5.6.b)$$

Proof of Theorem 5.6.3 By Theorem 5.6.2 the sets of propositions of Equations (5.6.a) and (5.6.b) can be assumed to be ideals. Concerning Equation (5.6.a), by Theorem 5.3.2, $(I \wedge J) \supset I$. Therefore by Theorem 5.6.1, $m(I \wedge J) \subset m(I)$. Similarly, $m(I \wedge J) \subset m(J)$. So $m(I \wedge J) \subset m(I) \cap m(J)$. Conversely, if $\omega \in m(I) \cap m(J)$, then ω satisfies all propositions of I and of J. Since $I \wedge J$ is formed by conjunctions of propositions of I with propositions of J, ω satisfies all the propositions of $(I \wedge J)$. That is, $\omega \in m(I \wedge J)$. This establishes Equation (5.6.a). Concerning equation (5.6.b), by Theorem 5.3.2, $I \vee J = I \cap J \subset I$. So by Theorem 5.6.1, $m(I \vee J) \supset m(I)$. By symmetry $m(I \vee J) \supset m(J)$. So $m(I \vee J) \supset m(I) \cup m(J)$. Conversely, if $\omega \in m(I \vee J)$, then ω satisfies all propositions of $I \vee J$. Now if ω doesn't satisfy some proposition p in I, then since ω satisfies every proposition in $\{p\} \vee J = [p \vee q: q \text{ in } J\}$, ω must satisfy every proposition in J. This establishes Equation (5.6.b). This completes the proof of Theorem 5.6.3.

The function m also preserves complements:

5.6.4 Theorem: The Models of ~p and the Models of p
If p is a proposition, then

$$m(\sim p) = (m(p))' \quad (5.6.c)$$

Proof of Theorem 5.6.4 By Definition 5.4.1, for any model $\omega \in \Omega$, either p is true in ω or else ~p is true in ω. That is, ω is in $m(p)$ or else ω is in $m(\sim p)$. That is, $(m(p))' = m(\sim p)$. This completes the proof of Theorem 5.6.4.

That the extension function m is one-to-one and so an isomorphism on ideals follows from K. Gödel's first-order completeness theorem [17]:

5.6.5 Theorem: The Extension Function is an Isomorphism
The extension function m is one-to-one from the ideals of L to the ideals of Ω. Thus m is an isomorphism.

Proof of Theorem 5.6.5 By the completeness theorem any proposition p that is true in every model of L is provable in L, which means p is in T, the ideal of theorems generated by the axioms. If m maps two ideals I and J onto the same ideal of models m(I) = m(J), then every proposition of I must be provable from J and the axioms A. That is, I ⊂ J ∧ A. Therefore, (I|A) ⊂ (J ∧ A | A) = (J|A). By symmetry (J|A) ⊂ (I|A). So I = (I|A) = (J|A) = J according to Definition 5.1.2.

The extension function m can now be extended to the conditional propositions and conditional ideals of L.

5.6.6 Definition: Models of Conditional Ideals
If p and q are propositions and I and J are ideals of L, then

$$m(q|p) = (m(q) | m(p)) \quad (5.6.d)$$

and

$$m(J|I) = (m(J) | m(I)), \quad (5.6.e)$$

where, in view of Theorem 5.6.2, outer parentheses indicating ideals have been removed from m(q), m(p), (m(J), and m(I).

5.6.7 Theorem: The Model Function m Is Well-Defined
The model function m is well defined by Definition 5.6.6. That is, if (J|I) = (K|I) then (m(J) | m(I)) = (m(K) | m(I)).

Proof of Theorem 5.6.7 Since (J|I) = (K|I), J ∧ I = K ∧ I. So m(J ∧ I) = m(K ∧ I). That is, m(J) ∩ m(I) = m(K) ∩ m(I). So m(J) | m(I) = m(K) | m(I).

Now that the isomorphism m has been extended to the conditional propositional ideals (J|I) onto the conditional models (m(J) | m(I)), it is clear that every logical expression p has an equivalent expression in the language of models (examples). See Figure 3.1.

5.7 Probability of an Arbitrary Conditional Proposition
Having defined the idea of a conditional logic L/I, it is now time to attach a (conditional) probability measure P to the collection Ω, of all models of L, and then to the conditional models of L.

5.7.1 Definition: A Probability Measure for a Boolean Logic

A probability measure P for a Boolean logic $\mathcal{L}$ with extension set Ω of models is a Kolmogorov probability measure (see Section 1.13) for which the collection of measurable subsets, $\mathcal{B}$, includes all the ideals of the form

$$\{m(I): I \text{ is an ideal of } L\} \qquad (5.7.a)$$

Actually, since there is a one-to-one correspondence between ideals of L and the ideals of Ω, Equation (5.7.a) is equivalent to requiring $\mathcal{B}$ to include all the ideals of Ω. The probability of an arbitrary proposition in L can now be defined:

5.7.2 Definition of the Probability of a Proposition and of an Ideal

If p is a proposition of a Boolean logic $\mathcal{L}$ and P is a probability measure on the collection of models, Ω, of $\mathcal{L}$, then P(p), the probability of p, is given by

$$P(p) = P(m(p)). \qquad (5.7.b)$$

Similarly, if I is an ideal of propositions of L, then

$$P(I) = P(m(I)). \qquad (5.7.c)$$

The probability of I is just the probability of the models (examples, cases, instances) in which the propositions of I are true.

Finally, the probability of conditional expressions such as (q|p) and (J|I) where (p) and I are ideals with positive probability can be defined:

$$P(q|p) = P(m(q|p)) \qquad (5.7.d)$$

and

$$P(J|I) = P(m(J|I)). \qquad (5.7.e)$$

As usual, $P(q|p) = P(q \wedge p) / P(p)$ and $P(J|I) = P(J \wedge I) / P(I)$. For instance, $P(m(q|p)) = P(m(q) \mid m(p)) = P(m(q) \cap m(p)) / P(m(p)) =$

$P(m(q \wedge p)) / P(m(p)) = P(q \wedge p) / P(p)$.

The model function m is an isomorphism between propositional ideals and model ideals, and the correspondence should be preserved when operating on conditionals:

5.8 The Model Function m is an Isomorphism on Ideals

Theorem If (q|p) and (s|r) are two conditional propositions, then

$$m((q|p) \vee (s|r)) = m(q|p) \cup m(s|r) \qquad (5.8.a)$$

and

$$m((q|p) \wedge (s|r)) = m(q|p) \cap m(s|r). \qquad (5.8.b)$$

Proof of Theorem: $m((q|p) \vee (s|r)) = m(qp \vee sr \mid p \vee r) = m(qp \vee sr) \mid m(p \vee r) = (m(qp) \cup m(sr) \mid m(p) \cup m(r)) = (m(qp) \mid m(p)) \cup ((m(sr) \mid m(r)) = m(q|p) \cup m(s|r)$. Similarly for conjunction. This completes the proof of Theorem 5.8.

It should not really come as a surprise that propositions that are not equivalent as conclusions may be equivalent as premises.

5.9 Statements Equivalent When Wholly True

5.9.1 Theorem: If Wholly True q ∈ (p), q ∨ ~p, & (q|p) are Equivalent

If p and q are propositions then "$q \in (p)$", "$q \vee \sim p$", and "(q|p)" are all always true if any one of them is always true.

Theorem 5.9.1 follows immediately from the next two lemmas.

5.9.2 Lemma: Equivalence of q ∈ (p) and (q ∨ ~p) if either is Certain

In the context of Theorem 5.9.1, $q \in (p)$ is always true if and only if $(q \vee \sim p)$ is always true.

Proof of Lemma 5.9.2 Suppose $q \in (p)$ is (always) true. Then $q = p \vee r$, for some r in L. Therefore, $q \vee \sim p = (p \vee r) \vee \sim p = (p \vee \sim p) \vee r = 1$. That is, $(q \vee \sim p)$ is (always) true. Conversely, suppose $q \vee \sim p$ is always true. That is, $q \vee \sim p = 1$. So $p = 1 \wedge p = (q \vee \sim p) \wedge p = (q \wedge p) \vee (\sim p \wedge p) = q \wedge p$. So $p \vee q = (q \wedge p) \vee q = q$. Therefore $q = p \vee r$ for some r, namely q, in L. This completes the proof of Lemma 5.9.2.

This lemma shows that being in the principal ideal generated by a proposition p is equivalent to being materially implied by p. The next lemma extends this equivalence to (q|p).

5.9.3 Lemma: Equivalence of q ∈ (p) and (q|p) if either is Certain

In the context of Theorem 5.9.1, q ∈ (p) is (always) true if and only if (q|p) is (always) true.

Proof of Lemma 5.9.3 Suppose q ∈ (p). Then q = q ∨ p and so p ∧ q = p ∧ (q ∨ p) = (p ∧ q) ∨ (p ∧ p) = p. So (q|p) = (p ∧ q | p) = (p|p) = (1|p). That is, (q|p) is always true. Conversely, if (q|p) = (1|p) then q ∧ p = p and so q ∨ p = q ∨ (q ∧ p) = (q ∨ q) ∧ (q ∨ p) = q. That is, q is in (p). This completes the proof of Lemma 5.9.3.

5.9.4 Corollary: p = q if and only if they generate the same ideal.

Although the three propositions of Theorem 5.9.1 are equivalent whenever any one of them is always true, it may come as a surprise to find that they all have generally different probabilities in case they are sometimes false — that is, if they are false in some models but true in others. q ∈ (p) can have a completely different probability distribution when considered merely sometime true. It is the probability that q is in the ideal (p).

5.10 The Probability of the Contrapositive of a Conditional

Theorem 5.10: If neither P(p) nor P(~q) is 0, then

$$P(\sim p \mid \sim q) = P(q|p) + [1 - P(p) / P(\sim q)] [1 - P(q|p)] \quad (5.10)$$

The obvious corollary is that P(~p | ~q) = P(q|p) if and only if P(q|p) =1 or P(p) = P(~q). That is, the contrapositive has the same probability as its corresponding implication when each has probability one or when their premises have equal probability, that is, when P(p) and P(~q) have the same probability, which means P(p) + P(q) = 1.

Proof of Theorem 5.10 By multiplying the two factors and simplifying, the right side of (5.10) becomes

$$P(q|p) + [1 - P(p) / P(\sim q)] [1 - P(q|p)]$$
$$= 1 - P(p)/P(\sim q) + [P(p)/P(\sim q)]P(q|p)$$
$$= 1 - [1 - P(q|p)] P(p)/P(\sim q)$$
$$= 1 - P(\sim q|p)P(p) / P(\sim q)$$
$$= 1 - P(\sim q \wedge p) / P(\sim q)$$
$$= 1 - P(p | \sim q)$$
$$= P(\sim p | \sim q)$$

Note that for the material conditional, there is never any difference between an implication and its contrapositive because $q \vee \sim p = \sim p \vee \sim(\sim q)$.

G. Boole realized that propositions that are equivalent when considered always true or always false may not be equivalent when considered only probable. Theorem 5.10 implies that a conditional proposition and its contrapositive are not equivalent when considered only probable.

5.11 The Probability of the Converse

The well-known theorem by Bayes [3, 4] supplies a relationship between the probability of a conditional implication (q|p) and the probability of its converse (p|q): If p and q are propositions, then, at least where defined,

$$P(p|q) = P(q|p)P(p) / [P(q|p)P(p) + P(q | \sim p)P(\sim p)] \quad (5.11)$$

5.12 The System of Boolean Fractions

Definition The set $\mathcal{L}/\mathcal{L}$ of a Boolean logic $\mathcal{L}$ is the set of *Boolean fractions*

$$L/L = \{(q|p): q, p \in L\} \quad (5.12)$$

under the 4 operations on conditional propositions of Section 3.10.

5.13 Formal Axioms of Conditional Probability Logic

5.13.1 Definition of a Conditional Probability Logic

A conditional probability logic (CPL) is a structure ($\mathcal{L}/\mathcal{L}$, $\mathcal{P}$) consisting of the set L/L of Boolean fractions of a Boolean propositional logic $\mathcal{L}$ together with a probability space $\mathcal{P} = (\Omega, \mathcal{B}, P)$, where Ω is some well-defined set of models of $\mathcal{L}$, and P is a probability measure on some collection $\mathcal{B}$ of subsets of Ω that includes {models m(I): I any subset of the set L of all propositions of $\mathcal{L}$}.

Since a set I of propositions and the ideal (I) generated by I have the same models, this is equivalent to defining P on all ideals of models of $\mathcal{L}$.

The structure of a conditional probability logic as just defined is sufficiently broad so as to include any Boolean propositional logic as well as any Kolmogorov probability space.

5.13.2 Theorem: Every Probability Space is a CPL

Every Kolmogorov probability space $\mathcal{P} = (\Omega, \mathcal{B}, P)$ is represented in a conditional probability logic ($\mathcal{L}/\mathcal{L}$, $\mathcal{P}$), where $\mathcal{L}$ is some Boolean propositional logic and Ω is the collection of all axiomatically allowed models of $\mathcal{L}$.

5.13.3 Theorem: Every Propositional Logic is a CPL

Every Boolean propositional logic $\mathcal{L}$ having at least one model is represented in a conditional probability logic ($\mathcal{L}/\mathcal{L}$, $\mathcal{P}$), where $\mathcal{P}$ is a probability space $(\Omega, \mathcal{B}, P)$ whose universe Ω is the nonempty set of all (axiomatically allowed) models of $\mathcal{L}$.

Proof of Theorem 5.13.2 Starting with a well-defined probability space $\mathcal{P} = (\Omega, \mathcal{B}, P)$, the required conditional probability logic ($\mathcal{L}/\mathcal{L}$, $\mathcal{P}$) can be formed by specifying as axioms of $\mathcal{L}$ the implicit or explicit description of the collection of all possible occurrences Ω of $\mathcal{L}$. This completes the proof of Theorem 5.13.2.

Proof of Theorem 5.13.3 Starting with a propositional logic $\mathcal{L}$, the set of all models Ω of $\mathcal{L}$ can be given a probability measure P defined upon some σ-algebra $\mathcal{B}$ of subsets of Ω that includes the collection {m(I): I a subset of L). This can be done by using the σ-algebraic closure of the collection {m(I):I a subset of L} as $\mathcal{B}$ and then defining a probability measure P whose domain is $\mathcal{B}$. There are many ways to assign probabilities. To show the existence of at least one such probability measure P, let ω be a model of L. For $B \in \mathcal{B}$, define P(B) to be 1 if ω is in B but 0 if ω is not in B. Then since ω is in Ω, $P(\Omega)=1$. Furthermore, $P(B) \geq 0$ for all B in $\mathcal{B}$. Finally, suppose B_1, B_2, B_3, are disjoint subsets in $\mathcal{B}$. If ω is in $(B_1 \cup B_2 \cup B_3 ...)$, then ω is in B_j for some integer j, and ω is not in B_i unless i = j. So $P(B_j) = 1$ and $P(B_j) = 0$ for i not equal to j. Therefore $(P(B_1 \cup B_2 \cup B_3 ...) = 1$ and $P(B_1) + P(B_2) + P(B_3) + ... = 1$. On the other hand, if ω is not in $(B_1 \cup B_2 \cup B_3 ...)$, then $P(B_1 \cup B_2 \cup B_3 ...) = 0$ and $P(B_1) + P(B_2) + ... = 0$. Therefore in any case $P(B_1 \cup B_2 \cup B_3 ...) = P(B_1) + P(B_2) + P(B_3) + ...$. Thus P is a probability measure on B. This completes the proof of Theorem 5.13.3.

In view of Theorems 5.13.2 and 5.13.3, all classical probability problems together with all Boolean propositional logics are incorporated in conditional probability logic.

5.14 Finite Conditional Probability Logics

The fact that complex conditionals can be reduced by the operations (4.8.a-d) to a single conditional having only Boolean components implies that finite Boolean logics $\mathcal{L}$ generate finite algebras $\mathcal{L}/\mathcal{L}$ of conditional propositions. These Boolean fractions - ordered pairs of Boolean propositions or events - are also called conditional events.

A finite Boolean logic is generated by a finite number of atomic propositions. (A nonzero proposition p is said to be atomic if and only if for all propositions $q \in L$ either $q \wedge p = p$ or $q \wedge p = 0$.)

It is helpful to examine the smallest closed systems of Boolean fractions on the way to considering all finite systems of Boolean fractions.

The smallest system of Boolean fractions $\mathcal{L}_1/\mathcal{L}_1$ is the 3-member set $\{(0|1), (1|1), (1|0)\}$. The first two entries are identified with the 0 and 1 of the initial 2-valued Boolean logic $\mathcal{L}_1 = \{0,1\}$, and $(1|0)$ is the "undefined" conditional U that is equivalent to any conditional with a false premise. Thus

$$\mathcal{L}_1/\mathcal{L}_1 = \{0, 1, U\}. \tag{5.14.a}$$

Equations of Section 2.0 extend the usual operations on 1 and 0. The next smallest system of Boolean fractions has nine members:

$$\mathcal{L}_2/\mathcal{L}_2 = \{0, 1, p, \sim p, (0|p), (1|p), (0|\sim p), (1|\sim p), U\} \tag{5.14.b}$$

where $\mathcal{L}_2 = \{1, 0, p, \sim p\}$ has two atoms, p and $\sim$p. There are twenty-seven members in the algebra of Boolean fractions generated by three (nonzero) atomic Boolean propositions p, q, and r:

$\mathcal{L}_2/\mathcal{L}_2 = \{0, 1, p, q, r, p \vee q, p \vee r, q \vee r, (0|p), (0|q), (0|r), (1|p),$
$(1|q), (1|r), U, (0 \mid p \vee q), (0 \mid p \vee r), (0 \mid q \vee r), (1 \mid p \vee q), (1 \mid p \vee r),$
$(1 \mid q \vee r), (p \mid p \vee q), (q \mid p \vee q), (p \mid p \vee r), (r \mid p \vee r), (q \mid q \vee r),$
$(r \mid q \vee r): p \vee q \vee r = 1, p \wedge q = p \wedge r = q \wedge r = 0\}$
(5.14.c)

It is well known [54] that any Boolean algebra $\mathcal{L}$ generated by a finite number N of atoms has 2^N elements, because it is isomorphic to the collection of all subsets of the N atoms under union, intersection, and complement. Each proposition $q \in L$ corresponds to an ideal subset (q) generated by the disjunction of one of the 2^N subsets of the N atoms. There are also 2^N ideals of L.

5.14.1 Theorem: The Number of Conditional Propositions in a CPL

There are 3^N distinct conditional propositions in the algebra of Boolean fractions generated by a Boolean logic with N atomic propositions.

Proof of Theorem 5.14.1 All propositions may be premises of conditionals. By Definition 5.3.4 two conditionals with nonequivalent premises are not equivalent, since their premises do not generate the same ideal. Therefore the number of nonequivalent conditional propositions in L/L is the sum of the numbers of nonequivalent conditionals in each L/I for all the nonequivalent ideals I of L.

The nonequivalent ideals of L are generated by the various different nonempty subsets of the N atoms of L. For each M < N there are N over M different M-element subsets of the N atoms that can be disjoined to generate a conditioning ideal I. That is,

There are $\binom{N}{M}$ different conditional logics L/I for which I is generated

by the disjunction of some M of the N atoms of L   (5.14.d)

The zero ideal (0) of all propositions generated by the proposition 0 is counted when M = 0.

For notational simplicity suppose that the first M atoms form the subset of N atoms. Then each member $(q \mid a_1 \vee a_2 \ldots \vee a_M)$ of $L/(a_1 \vee a_2 \ldots \vee a_M)$ is equivalent to its reduced form $(q \wedge (a_1 \vee a_2 \ldots \vee a_M) \mid (a_1 \vee a_2 \ldots \vee a_M)) = (q \wedge a_1) \vee (q \wedge a_2) \vee \ldots \vee (q \wedge a_M) \mid (a_1 \vee a_2 \ldots \vee a_M)$. Since the a's are atoms, each conjunction $q \wedge a_i$ is either a_i or 0, depending on whether the development of q as a disjunction of the atoms of L includes a_i or not. Atoms that are not among the M atoms of I are equivalent to 0 in $L/(a_1 \vee a_2 \ldots \vee a_M)$. Each of the 2^M subsets of the M atoms can be disjoined to form a nonequivalent member of $L/(a_1 \vee a_2 \ldots \vee a_M)$, because each when conjoined with $(a_1 \vee a_2 \ldots \vee a_M)$ will result in the disjunction of a different subset of the m atoms. But there are exactly 2^M different subsets of the M atoms. Therefore,

There are exactly 2^M different elements in a conditional logic L/I whose condition is generated by the disjunction of some M of the N atoms of L.   (5.14.e)

Combining equations (5.14.d) and (5.14.e) and summing over M from 0 to N atoms, yields for the number C_N of distinct elements in $\mathcal{L}_N/\mathcal{L}_N$,

$$C_N = \binom{N}{0}2^0 + \binom{N}{1}2^1 + \ldots + \binom{N}{N}2^N = (1+2)^N = 3^N \qquad (5.14.f)$$

5.15 Some Boolean Properties No Longer True

It is easy to see that commutativity, associativity, and idempotency still hold in the system of Boolean fractions. But there is a partial loss of distributivity, absolute complements, and units:

$$p \vee [1 \wedge (0|p)] \neq [p \vee 1] \wedge [p \vee (0|p)] \text{ unless } p = 1 \qquad (5.15.a)$$

$$p \wedge [0 \wedge (1|\sim p)] \neq [p \wedge 0] \vee [p \wedge (1|\sim p)] \text{ unless } p = 0 \quad (5.15.b)$$

$$(q|p) \vee 1 = 1; \text{ but } (q|p) \wedge 1 = q \vee \sim p \quad (5.15.c)$$
$$(q|p) \wedge 0 = 0; \text{ but } (q|p) \vee 0 = q \wedge p \quad (5.15.d)$$
$$(q|p) \wedge \sim(q|p) = (0|p), \text{ which is not } 0 \text{ unless } p = 1 \quad (5.15.e)$$
$$(q|p) \vee \sim(q|p) = (1|p), \text{ which is not } 1 \text{ unless } p = 1 \quad (5.15.f)$$

The De Morgan formulas still hold however with respect to the relative complements.

These ordered pairs of propositions with their four extended operations, also called conditional events, conditional propositions, or simply conditionals, provide the basis for attaching consistent conditional probabilities to any conditional proposition and to combinations of conditionals when operated upon by the four operations "and", "or", "not" and "if".

5.16 Standard Logical Formulas and Probabilities

Many formulas familiar from probability theory have logic counterparts:

$$P(q \wedge p) = P(q|p)P(p) \text{ and } (q \wedge p) = (q|p) \wedge p; \quad (5.16.a)$$

$$P(r|p) = P(q \wedge r | p) + P(\sim q \wedge r | p), \text{ and} \quad (5.16.b)$$
$$(r|p) = (q \wedge r | p) \vee (\sim q \wedge r | p); \quad (5.16.c)$$

$$P(r|p) = P(q|p)P(r | p \wedge q) + P(\sim q|p)P(r | p \wedge \sim q), \text{ and} \quad (5.16.d)$$
$$(r|p) = (q|p) \wedge (r | p \wedge q) \vee (\sim q|p) \wedge (r | p \wedge \sim q); \quad (5.16.e)$$

$$P(r \wedge q | p) = P(q|p)P(r | p \wedge q), \text{ and} \quad (5.16.f)$$
$$(r \wedge q | p) = (q|p) \wedge (r | p \wedge q); \quad (5.16.g)$$

$$P(r | p \wedge q) = P(q \wedge r | p) / P(q|p), \text{ and} \quad (5.16.h)$$
$$(r | p \wedge q) = ((q \wedge r | p) | (q|p)); \quad (5.16.i)$$

$$P(r | p \wedge q) = P(r|q) / P(p|q), \text{ and} \quad (5.16.j)$$
$$(r | p \wedge q) = ((r|q) | (p|q)). \quad (5.16.k)$$

Modus ponens [if p is true and "if p implies q" then q is true] can be proved in the object language:

$$(q \mid [p \wedge (q|p)]) = (q \mid p \wedge q) = (1 \mid p \wedge q) \qquad (5.16.1)$$

That is, q is certain (always true) given p and "q given p".

5.17 Nonstandard Formulas & Associated Probabilities

The conjunction of a conditional with its converse is equivalent in reduced form to a simple conditional, namely, to the conjunction of the premise and the conclusion given the disjunction of the premise and the conclusion. Its conditional probability is therefore the ratio of the probabilities of the conjunction to the disjunction of the two propositions in question:

$$(q|p) \wedge (p|q) = (p \wedge q \mid p \vee q) \qquad (5.17.a)$$

and

$$P[(q|p) \wedge (p|q)] = P(p \wedge q \mid p \vee q) \qquad (5.17.b)$$

One of the more noteworthy qualities of the conditional logic is that two elements can have the same disjunction and conjunction without being themselves the same:

$$(q|p) \vee (q|\sim p) = q = (q|p) \wedge (q|\sim p) \qquad (5.17.c)$$

and

$$P[(q|p) \vee (q|\sim p)] = P(q) = P[(q|p) \wedge (q|\sim p)] \qquad (5.17.d)$$

but (q|p) is not (q|~p). Note that in Boolean logic this cannot happen. For if $p \wedge q = p \vee q$ then $p = p \wedge 1 = p \wedge (q \vee 1) = (p \wedge q) \vee (p \wedge 1) = (p \wedge q) \vee p = (p \vee q) \vee p = p \vee q$. By symmetry $q = q \vee p$. Therefore $p = q$.

Earlier, equation 5.17.c was used to characterize the $\wedge$ and $\vee$ operations on conditionals. They are axiomatic because they represent the method of proof by breaking into disjoint cases. To prove q it suffices to prove "q in case p and q in case not p".

The following formula expresses $p \wedge q$ in terms of (q|p) and (0|~p):

$$(q|p) \vee (0|\sim p) = p \wedge q = (q|p) \wedge (0|\sim p) \qquad (5.17.e)$$

The next equation expresses (q ∨ ~p) in terms of (q|p) and (1|~p):

$$(q|p) \vee (1 \,|\, {\sim}p) = q \vee {\sim}p = (q|p) \wedge (1 \,|\, {\sim}p) \qquad (5.17.f)$$

The following equation is intuitively clear:

$$(q|p) \vee (q|r) = (q \,|\, p \vee r) \qquad (5.17.g)$$

Whenever at least one side of these equations is a simple conditional, then both sides have a well-defined and equal conditional probability.

5.18 Canonical Form of (q|p) ∨ (s|r) and (q|p) ∧ (s|r)

Canonical development of the reduced conclusion in terms of the four propositions p, q, r, and s yields five potential atoms for the conjunction of two conditionals:

$$(q|p) \wedge (s|r) = [(q \wedge p \wedge s \wedge r) \vee (q \wedge {\sim}p \wedge s \wedge r) \vee (q \wedge p \wedge s \wedge {\sim}r)$$
$$\vee ({\sim}q \wedge {\sim}p \wedge s \wedge r) \vee (q \wedge p \wedge {\sim}s \wedge {\sim}r) \,|\, (p \vee r)] \qquad (5.18.a)$$

For the disjunction of these same two conditionals, two more atomic conditionals appear in the canonical development:

$$(q|p) \vee (s|r) = [(q|p) \wedge (s|r)] \vee [({\sim}q \wedge p \wedge s \wedge r) \vee (q \wedge p \wedge {\sim}s \wedge r) \,|\, (p \vee\ r)]$$
$$(5.18.b)$$

See Figure 5.2 for a Venn diagram illustrating these subsets and conditional subsets.

$m((q|p) \wedge (s|r)) = (1 \cup 2 \cup 3 \cup 4 \cup 5 \mid m(p \vee r))$
$m((q|p) \vee (s|r)) = (1 \cup 2 \cup 3 \cup 4 \cup 5 \cup 6 \cup 7 \mid m(p \vee r))$

1. $m(q \wedge p \wedge s \wedge r)$
2. $m(q \wedge \sim p \wedge s \wedge r)$
3. $m(q \wedge p \wedge s \wedge \sim r)$
4. $m(\sim q \wedge \sim p \wedge s \wedge r)$
5. $m(q \wedge p \wedge \sim s \wedge \sim r)$
6. $m(\sim q \wedge p \wedge s \wedge r)$
7. $m(q \wedge p \wedge \sim s \wedge r)$
8. $m(\sim q \wedge p \wedge s \wedge \sim r)$
9. $m(q \wedge \sim p \wedge \sim s \wedge r)$

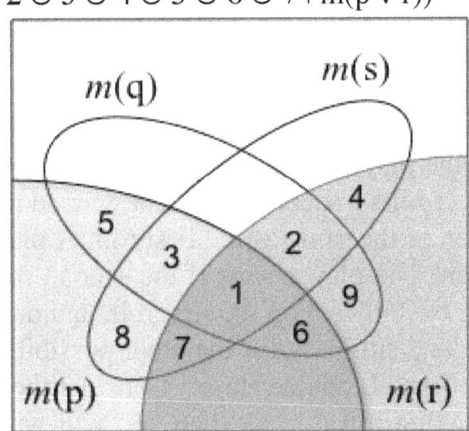

Figure 5.2 Development of Conjunctions and Disjunctions of Conditionals

From Equations (5.18.a) and (5.18.b) it follows that

$(q|p) \vee (s|r) = (q|p) \wedge (s|r)$ if and only if
$$p \wedge r \wedge [(\sim q \wedge s) \vee (q \wedge \sim s)] = 0. \quad (5.18.c)$$

That is, $(q|p)$ and $(s|r)$ have the same (equivalent) disjunction as conjunction if and only if there is never a case when both premises are true but just one conclusion is true. In reduced form,

$(q|p) \wedge (s|r) = [(q \wedge p \wedge \sim r) \vee (q \wedge p \wedge s \wedge r) \vee (\sim p \wedge s \wedge r) \mid (p \vee r)]$

$(q|p) \vee (s|r) = [(q \wedge p) \vee (s \wedge r) \mid (p \vee r)] \quad (5.18.d)$

The right hand sides of these equations can be written more concisely as follows:

$(q|p) \wedge (s|r) = (qpr' \vee qpsr \vee p'sr \mid p \vee r)$
$(q|p) \vee (s|r) = (qp \vee sr \mid p \vee r) \quad (5.18.e)$

5.19 A Sample Calculation

A dice game makes a good context in which to consider some simple applications of conditional probability logic. The rules (axioms) of the game are as follows. Two six-sided cubes (dice) have their faces numbered 1 through 6. A player rolls the dice up against a wall, and when they come to rest, the sum of the two numbers facing up is noted. If the sum is 2, 3, or 12, the player loses immediately. If the sum is 7 or 11, the player wins immediately. If the sum is 4, 5, 6, 8, 9, or 10, the player rolls a second time. The second roll does not count unless it is a seven or the same as the first roll. A player loses if his second roll is a seven. The player wins if the second roll is the same as the first. Denote by "6F" a 6 on the first roll, and denote by "7S" a 7 on the second roll, and similarly for the other possibilities in each roll. Let W denote "win" and L denote "lose". Consider the following proposition: If your first roll is a six then you will win, or if your first roll is a four then you will lose. In symbols this is

$$(W|6F) \vee (L|4F) \qquad (5.19)$$

What simple conditional is equivalent to the compound conditional proposition of (5.19)? What is the conditional probability of (5.19)? Is (5.19) more true than false? If so, how much?

The expression can be simplified and its probability determined as follows:

$$(W|6F) \vee (L|4F) = ((W \wedge 6F) \vee (L \wedge 4F) \mid 6F \vee 4F)$$
$$= (W \wedge 6F \mid 6F \vee 4F) \vee (L \wedge 4F \mid 6F \vee 4F)$$

Since winning with six first is disjoint from losing with four first,

$$P((W|6F) \vee (L|4F)) = P(W \wedge 6F \mid 6F \vee 4F) + P(L \wedge 4F \mid 6F \vee 4F)$$
$$= P(W \wedge 6F) / P(6F \vee 4F) + P(L \wedge 4F) / P(6F \vee 4F)$$

Now $P(6F \vee 4F) = 5/36 + 3/36$. It remains to determine $P(W \wedge 6F)$ and $P(L \wedge 4F)$. $P(W \wedge 6F) = P(W|6F)P(6F)$. Given 6F, W is equivalent to 6S. So $(W|6F) = (6S|6F)$. But $(6S|6F)$ is equivalent to $(6S|7S \vee 6S)$ because as a condition, 6F is equivalent to $(7S \vee 6S)$. Therefore $(W|6F) = (6S|6F) = (6S|7S \vee 6S)$. So $P(W|6F) = P(6S|7S \vee 6S) = P(6S)/P(7S \vee 6S) = (5/36) / (6/36 + 5/36) = 5/11$. Similarly for P(L $\wedge$

4F) = P(L|4F)P(4F), where P(L|4F) = P(7S | 7S ∨ 4S) = 6/9. All probabilities are now known. Substitution yields a final value of 47/88. So the statement is more true than false, exactly 3/88 better than an even bet.

> The set of all *consequences of the set A* is the intersection of all sets which contain the set A and are closed under the given rules of inference.
> Alfred Tarski [56] 'Über einige fundamentale Begriffe der Metamathematik', 1930

Chapter 6

Deduction of Conditionals by Conditionals

> "There is one important respect, however, in which quasi-conjunction differs from ordinary conjunction: When its parts are not factual the quasi-conjunction may not p-entail them."
> E. Adams [153], *A Primer of Probability Logic*, 1998, p165.

6.0 Varieties of Deduction Premised on 3-Valued Propositions

If statements of the certainty of a proposition or conditional proposition are allowed such as "c is never false" or "(c|d) is never false", then there are many more forms of deduction involving the equivalence relation "=". It becomes clear that deduction and inference involving conditionals requires special care in specifying exactly what is being assumed (the conditions), and secondly exactly what is being deduced or inferred, and thirdly exactly what type of deduction between conditionals is being invoked. (See Calabrese [69].)

6.1 Deductive Relations and Deductively Closed Sets of Conditionals

Due to the existence of four (not just two) propositions between two conditionals, deduction takes several forms. To put them into a common context the following algebraic definitions are useful:

6.1.1 Definition: Deductive Relation

A *deductive relation* (also called a *preorder* or an *implication*) is a reflexive and transitive relation, $\leq$, on a set.

6.1.2 Definition: Boolean Extension Property

A deductive relation $\leq$ for conditionals has the *Boolean Extension Property* if and only if for all propositions a, b, and c,

$$ab \leq cb \quad \text{implies} \quad (a|b) \leq (c|b).$$

That is, a deductive relation $\leq$ on $\mathcal{B}/\mathcal{B}$ has the Boolean Extension Property if it extends the Boolean deduction relation $\leq$ of every Boolean sub-algebra $\mathcal{B}/b$, of $\mathcal{B}/\mathcal{B}$, that is, if it agrees with Boolean deduction on every Boolean sub-algebra $\mathcal{B}/b$, of $\mathcal{B}/\mathcal{B}$.

6.1.3 Definition: Well-Defined Deductive Relations

The deductive relation $\leq$ is *well-defined* with respect to equality (=) of conditionals if and only if whenever $(a|b) = (a_1|b_1)$ and $(c|d) = (c_1|d_1)$ and $(a|b) \leq (c|d)$ then $(a_1|b_1) \leq (c_1|d_1)$.

6.1.4 Theorem: Boolean Extension Property Implies Well-Defined

If a deductive relation $\leq$ has the Boolean Extension Property then it is well-defined with respect to equality (=) of conditionals.

Proof of Theorem 6.1.4: Suppose $(a|b) = (a_1|b_1)$ and $(c|d) = (c_1|d_1)$ and $(a|b) \leq (c|d)$. By the definition of equality (=) of conditionals, $b = b_1$ and $d = d_1$, and $ab = a_1b_1 = a_1b$ and similarly $cd = c_1d$. Since $\leq$ has the Boolean Extension Property and $a_1b \leq ab$, therefore $(a_1|b) \leq (a|b)$. So $(a_1|b_1) = (a_1|b) \leq (a|b) \leq (c|d) = (c|d_1) \leq (c_1|d_1)$ using both equivalence of conditionals and the Boolean Extension Property twice. By transitivity of $\leq$ it follows that $(a_1|b_1) \leq (c_1|d_1)$. That completes the proof of Theorem 6.14.

According to one standard definition a subset S of *unconditioned* events (or propositions) is a deductively closed set of events (or propositions) provided that 1) the conjunction of any two propositions in S is also in S, and 2) any event that subsumes an event in S is also in S. (A deductively closed set is a sum ideal as previously defined.) A similar definition also works for deductively closed subsets of conditional propositions (or conditional events) with respect to some specified deductive relation (preorder):

6.1.5 Definition: Deductively Closed Sets of Conditional Propositions

A subset H of $\mathcal{B}/\mathcal{B}$ is said to be *deductively closed* with respect to a deductive relation $\leq_x$ provided H has both of the following properties:

1) If $(a|b) \in H$ and $(c|d) \in H$ then $(a|b) \wedge (c|d) \in H$, and
2) If $(a|b) \in H$ and $(a|b) \leq_x (c|d)$ then $(c|d) \in H$.

The first property will be called *conjunctive closure* and the second will be called *deductive closure*. H is said to be a *deductively closed set* (DCS) of conditionals with respect to $\leq_x$. For Boolean propositions, $\leq_x$ can be the standard deduction relation.

6.1.6 Definition: Deductive Equivalence ($=_x$) of Conditionals

$(a|b) =_x (c|d)$ means $(a|b) \leq_x (c|d)$ and $(c|d) \leq_x (a|b)$.

6.2 Extensions of Boolean Implication

The following definitions are natural since they are equivalent in Boolean algebra but not so in the general algebra of conditionals.

6.2.1 Definition: Conjunctive Implication ($\leq_\wedge$)

$(a|b) \leq_\wedge (c|d)$ means $(a|b) \wedge (c|d) = (a|b)$

6.2.2 Definition: Disjunctive Implication ($\leq_\vee$)

$(a|b) \leq_\vee (c|d)$ means $(a|b) \vee (c|d) = (c|d)$

6.2.3 Definition: Probabilistically Monotonic Implication ($\leq_{pm}$)

$(a|b) \leq_{pm} (c|d)$ means $(c|d) \vee (a|b)' = (1 | d \vee b)$

Proving reflexivity and transitivity for the above three deductive relations is trivial except perhaps transitivity for the last one, which will be proved as a corollary to Theorem 6.2.9. Similarly, showing the following three plausible implication relations are reflexive and transitive is an easy corollary of Theorems 6.2.7 and 6.2.8, which together prove that the three are equivalent to one another.

6.2.4 Definition: Non-Falsity Implication ($\leq_{nf}$)

$$(a|b) \leq_{nf} (c|d) \quad \text{means} \quad (a \vee b') \leq (c \vee d')$$

That is, if (a|b) is not false then (c|d) is not false

6.2.5 Definition: Necessary Implication ($\leq_n$)

$$(a|b) \leq_n (c|d) \quad \text{means} \quad \text{if } (a|b) = (1|b) \text{ then } (c|d) = (1|d)$$

That is, if a is necessary given b then c is necessary given d.

6.2.6 Definition: Conditional Necessity Implication ($\leq_c$)

$$(a|b) \leq_c (c|d) \quad \text{means} \quad (c|d) \mid (a|b) = (1|d) \mid (a|b)$$

That is, given (a|b), c is necessary given d.

6.2.7 Theorem: Non-Falsity Implication Equals Conditional Necessity

Non-Falsity Implication is equivalent to Conditional Necessity Implication.

Proof of Theorem 6.2.7: (c|d) | (a|b) = (1|d) | (a|b) iff [(c|d) = (1|d) | (a|b)] iff (d ≤ c) | (a|b) iff (d | (a|b)) ≤ (c | (a|b)) iff d(a ∨ b') ≤ c(a ∨ b') iff dc'(a ∨ b') ≤ 0 iff (a ∨ b') ≤ (c ∨ d'). (The result also follows by simplifying (c|d) | (a|b) to (c | d(a ∨ b')) and (1|d) | (a|b) to (1 | d(a ∨ b')), which are equal. So cd(a ∨ b') = d(a ∨ b'), which is equivalent to (a ∨ b') ≤ (c ∨ d').)

6.2.8 Theorem: Non-Falsity Implication Equals Necessary Implication

Non-Falsity Implication is equivalent to Necessary Implication.

Proof of Theorem 6.2.8: $(a|b) = (1|b)$ means $ab = b$, which means $b \leq a$, which means $(a \vee b') = 1$. Similarly $(c|d) = (1|d)$ is equivalent to $(c \vee d') = 1$. So the Necessary Implication $(a|b) \leq_n (c|d)$ means that $(a \vee b') = 1$ implies $(c \vee d') = 1$, which can only be true providing $a \vee b' \leq c \vee d'$. (See the next theorem.

6.2.9 Theorem: The Implication of the Certainty of Non-Falsity

If whenever $a \vee b' = 1$ then $c \vee d' = 1$, then $a \vee b' \leq c \vee d'$. (The converse is trivial.)

Proof of Theorem 6.2.9: Suppose e is an arbitrary proposition. By hypothesis, if $(a \vee b' \mid e) = (1 \mid e)$ then $(c \vee d' \mid e) = (1 \mid e)$. So for all propositions e, $[(a \vee b') e = e]$ implies $[(c \vee d') e = e]$. That is, for all propositions e, $[e \leq (a \vee b')]$ implies $[e \leq (c \vee d')]$. Setting $e = (a \vee b')$ yields that $[(a \vee b') \leq (a \vee b')]$ implies $[(a \vee b') \leq (c \vee d')]$. Since $[(a \vee b') \leq (a \vee b')]$ is always true, so too must $[(a \vee b') \leq (c \vee d')]$.

Since these three relations are equivalent, it suffices to show any one of them to be reflexive and transitive. The non-falsity relation of Definition 6.2.4 is obviously reflexive and transitive.

6.2.10 Theorem: Deductive Relations Reduced to Boolean Relations

The deductive relations $\leq_\wedge$, $\leq_\vee$, $\leq_{pm}$, and $\leq_{nf}$ on conditionals defined by the following equations can be reduced to the Boolean relations listed on the right.

$\leq_\wedge$: $(a|b) \wedge (c|d) = (a|b)$ if & only if $(a \vee b') \leq (c \vee d')$ and $(b' \leq d')$
$\leq_\vee$: $(a|b) \vee (c|d) = (c|d)$ if & only if $(a \wedge b) \leq (c \wedge d)$ and $(b \leq d)$
$\leq_{pm}$: $(a|b)' \vee (c|d) = (1 \mid b \vee d)$ if & only if $(a \vee b') \leq (c \vee d')$ and $(a \wedge b) \leq (c \wedge d)$
$\leq_{nf}$: $[(c|d) \mid (a|b)] = [(1|d) \mid (a|b)]$ if & only if $(a \vee b') \leq (c \vee d')$

Proof of Theorem 6.2.10. Concerning the first equation of the theorem, suppose $(a|b) \wedge (c|d) = (a|b)$. Applying the conjunction operation and the definition of equivalence for conditionals yields the two equalities $abd' \vee b'cd \vee abcd = ab$ and $b \vee d = b$. So immediately from just the second equality it follows that $d \leq b$, which is equivalent to $b' \leq d'$. Since $b'd = 0$ the first equality becomes $abd' \vee 0 \vee abcd = ab$, which is equivalent to $ab(d' \vee cd) = ab$. The latter is equivalent to $ab \leq (d' \vee cd)$, and since $b' \leq d'$, it follows immediately that $b' \vee ab \leq (d' \vee cd)$, which is $(a \vee b') \leq (c \vee d')$. Reversing these steps produces the converse. For the second equality, applying the disjunction operation and the definition of equality of conditionals yields that $ab \vee cd = cd$ and $b \vee d = d$, which are equivalent to $ab \leq cd$ and $b \leq d$ respectively. Reversing these steps yields the converse of the second equation. Concerning the third equality of the theorem, applying negation and disjunction for conditionals and the definition of equivalence of conditionals yields $a'b \vee cd = b \vee d$ and $b \vee d = b \vee d$. Conjunction of both sides of the first equality by ab yields $abcd = ab$, which is equivalent to $ab \leq cd$. Conjunction of both sides of the first equality instead by $c'd$ yields $(a'b)(c'd) = (c'd)b \vee c'd = c'd$. So $c'd \leq a'b$, which, by taking complements of both sides and reversing the inequality, is equivalent to $(a \vee b') \leq (c \vee d')$. The converse of the third equality of the theorem follows since if $ab \leq cd$ and $c'd \leq a'b$ then $a'b \vee cd = (a'b \vee c'd) \vee (cd \vee ab) = (a'b \vee ab) \vee (cd \vee c'd) = b \vee d$. That is $a'b \vee cd = b \vee d$, which is equivalent to $(a|b)' \vee (c|d) = (1 \mid b \vee d)$. Finally, concerning the fourth equality of the theorem, applying the conditioning operation and the definition of equivalence of conditionals yields that $cd(a \vee b') = d(a \vee b')$. Disjunction on both sides of the latter equality by $d'(a \vee b')$ yields $(cd \vee d')(a \vee b') = (a \vee b')$, which is equivalent to $(a \vee b') \leq (c \vee d')$. Conversely, if $(a \vee b') \leq (c \vee d')$ then conjunction of both sides by d yields that $(a \vee b')d \leq cd$. So $(cd)(a \vee b')d = (a \vee b')d$. That is $(cd)(a \vee b') = (a \vee b')d$, which is equivalent to the left side of the fourth equality of the theorem. That completes the proof of Theorem 6.2.10.

The reduction of the relations listed in Theorem 6.2.10 to their associated Boolean relations also exhibits the obvious transitivity of those relations.

This theorem suggests that the Boolean relations on the right hand sides of "if & only if" in Theorem 6.2.10 define *elementary implica-*

tions on conditionals. Indeed, there is the following hierarchy of implications:

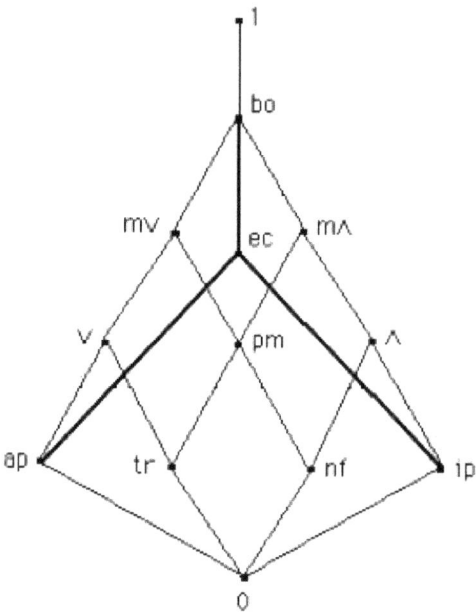

Figure 6.1 Hierarchy of Implications (Deductive Relations) for Conditionals

Elementary Implications

tr - Implication of Truth ($\leq_{tr}$)
  (a|b) $\leq_{tr}$ (c|d) iff $ab \leq cd$

nf - Implication of Non-Falsity ($\leq_{nf}$)
  (a|b) $\leq_{nf}$ (c|d) iff $(a \vee b') \leq_{nf} (c \vee d')$

ap - Implication of Applicability ($\leq_{ap}$)
  (a|b) $\leq_{ap}$ (c|d) iff $b \leq d$

ip - Implication of Inapplicability ($\leq_{ip}$)
  (a|b) $\leq_{ip}$ (c|d) iff $d \leq b$

Three Elementaries Combined

mv - (Probabilistically) Monotonic and Applicability Implication ($\leq_{mv}$)

m∧ - (Probabilistically) Monotonic and Inapplicability Implication ($\leq_{m\wedge}$)

Two Elementaries Combined

∨ - Disjunctive Implication ($\leq_\vee$)

pm - Probabilistically Monotonic Implication ($\leq_{pm}$)

∧ - Conjunctive Implication ($\leq_\wedge$)

ec - Implication of Equal Conditions ($\leq_{ec}$) (a|b) $\leq_{ec}$ (c|d) iff $b = d$

Trivial Implications

1 - Implication of Identity ($\leq_1$)
  (a|b) $\leq_1$ (c|d) iff (a|b) = (c|d)

0 - Universal Implication
  (a|b) $\leq_0$ (c|d) for all (a|b) & (c|d)

Boolean Deduction Extended

bo - Boolean Deduction ($\leq_{bo}$)
  (a|b) $\leq_{bo}$ (c|d) iff $b = d$ and $ab \leq cd$

Note also that all four elementary preorders have the Boolean Extension Property because for x ∈ {ap, ip}, the relation (a|b) $\leq_x$ (c|b) holds whether or not (ab ≤ cb) holds. For x = tr, (a|b) $\leq_x$ (c|b) reduces to the hypothesis ab ≤ cb, and finally for x = nf, (a|b) $\leq_x$ (c|b) reduces to a ∨ b' ≤ c ∨ b', which becomes ab ≤ cb after conjunction on both sides by b. Since ab ≤ cb implies (a|b) $\leq_x$ (c|b) for all four elementary preorders x, the other preorders in the hierarchy also satisfy the Boolean Extension Property. (See also Calabrese [64] 687-8 and [65] 85-100.)

6.2.11 Probability Relationships

Probability relationships naturally flow from the above implication relations, although among the 13 defined above only monotonic implication, and those above it in the hierarchy, ensure probabilistic monotonicity. That is, if (q|p) $\leq_{pm}$ (s|r) then P(q|p) ≤ P(s|r). For instance, non-falsity implication, (q|p) $\leq_{nf}$ (s|r), ensures only that P(q ∨ p') ≤ P(s ∨ r').

6.2.12 Conditional Equivalence

Although the three equivalent implication relations {$\leq_{nf}$, $\leq_c$, $\leq_n$} are reflexive and transitive, they are not anti-symmetric. That is, (a|b) $\leq_{nf}$ (c|d) and (c|d) $\leq_{nf}$ (a|b) together do not imply that (a|b) = (c|d). That is, (a ∨ b') = (c ∨ d') does not imply that b=d and ab = cd. As such, $\leq_{nf}$ is a quasi-order (also called a preorder), but not a partial order. Equivalently, (a|b) $\leq_{nf}$ (c|d) if and only if c'd ≤ a'b. The latter means that "if (c|d) is false then (a|b) is false." In these terms, (a|b) $=_{nf}$ (c|d) means that (a|b) is false if and only if (c|d) is false.

6.2.13 Conditional Implication and the Contrapositive

Note that a conditional proposition (a|b) and its contrapositive, (b'|a'), are non-falsely equivalent: (a|b) $=_{nf}$ (b'|a'). This is reassuring since a conditional proposition and its contrapositive should be logically equivalent when regarded as wholly non-false (as when assumed or conditioned upon) but not equivalent, nor even have the same probability, when regarded as partially false. In fact, a conditional (a|b) is false if and only if its contrapositive (b'|a') is false, but if either (a|b) or (b'|a') is true the other is inapplicable. They can also both be inapplicable. (See Calabrese [63] p222, for a comparison of the probability of (a|b) with the probability of (b'|a').)

6.2.14 Certainty Theorem
If, whenever ab = 1 then cd = 1, then ab ≤ cd. That is, if, whenever (a|b) is true then (c|d) is true, then ab ≤ cd. (The converse is easy.)

Proof of Theorem 6.2.14: If for all propositions e, [(ab | e) = (1 | e)] implies (cd | e) = (1 | e), then it follows as above that for all e, [e ≤ ab] implies [e ≤ cd]. Then setting e = ab yields that ab ≤ cd. That completes the proof of Theorem 6.2.14.

Thus the "necessarily true" preorder $\leq_{nt}$, defined by:

$$(a|b) \leq_{nt} (c|d) \quad \text{means} \quad \text{if } (ab = 1) \text{ then } (cd = 1)$$

is equivalent to the preorder $\leq_{tr}$, defined by (ab ≤ cd).

The deductive relations above and some new ones built up from the elementary ones have been organized into a hierarchy (Figure 6.1) defined completely in terms of Boolean relations on the Boolean components of the conditionals. These results are summarized in the following:

6.2.15 Interpretations of Elementary Deductive Conditionals

$\leq_{tr}$: ab ≤ cd   "If (a | b) is true then (c | d) is true",
$\leq_{nf}$: a ∨ b' ≤ c ∨ d'   "If (a | b) is not false then (c | d) is not false"
$\leq_{ap}$: b ≤ d   "If (a | b) is applicable then (c | d) is applicable"
$\leq_{ip}$: b' ≤ d'   "If (a | b) is inapplicable then (c | d) is inapplicable"

6.2.16 Two Elementary Deductive Relations Combined
From these four deductive relations, three of the previously identified deductive relations ($\leq_\wedge$, $\leq_\vee$, and $\leq_{pm}$) and a 4th can be defined by combining the properties of two of the elementary deductive relations:

$$(a|b) \leq_\wedge (c|d) \quad \text{if and only if} \quad (a \vee b') \leq (c \vee d') \text{ and } (b' \leq d')$$
$$\text{(Conjunctive Implication)}$$

$(a|b) \leq_v (c|d)$ if and only if $ab \leq cd$ and $b \leq d$
(Disjunctive Implication)

$(a|b) \leq_{pm} (c|d)$ if and only if $ab \leq cd$ and $(a \vee b') \leq (c \vee d')$
(Probabilistically Monotonic Implication)

$(a|b) \leq_{ec} (c|d)$ if and only if $b \leq d$ and $b' \leq d'$ (i.e. $b = d$)
(Equal Conditions Implication)

6.2.17 Three Elementary Deductive Relations Combined
Three additional deductive relations arise by combining the properties of three of the elementary deductive relations:

$(a|b) \leq_{m\wedge} (c|d)$ if and only if $ab \leq cd$, $(a \vee b') \leq (c \vee d')$, & $b' \leq d'$
$(a|b) \leq_{mv} (c|d)$ if and only if $ab \leq cd$, $(a \vee b') \leq (c \vee d')$, & $b \leq d$
$(a|b) \leq_{bo} (c|d)$ if and only if $b = d$ and either $ab \leq cd$ or $(a \vee b') \leq (c \vee d')$

Boolean deduction $\leq_{bo}$ results by combining $\leq_{ap}$ and $\leq_{ip}$ with either $\leq_{tr}$ or $\leq_{nf}$. Note that when $b = d$, the two conditionals $(a|b)$ and $(c|d)$ reside in the Boolean subalgebra $(\mathcal{B}|b)$, where Boolean deduction is determined by whether $ab \leq cb$.

6.2.18 Deductive Equivalence Theorem
For $x \in \{ap, tr, nf, ip, ec, 0\}$, $(a|b) =_x (c|d)$ does not imply $(a|b) = (c|d)$, but for the other seven deductive relations in the hierarchy, $(a|b) =_x (c|d)$ implies equality of $(a|b)$ and $(c|d)$ as conditionals.

Proof of Theorem 6.2.18 For $x=0$, $(a|b) =_x (c|d)$ is true for any two unequal conditionals. So equality of $(a|b)$ & $(c|d)$ is not implied. For $x \in \{ap, ip, ec\}$, $(a|b) =_x (c|d)$ simply implies that $(a|b)$ and $(c|d)$ have common condition. That is, $b = d$, but they need not be equal. For $x = nf$, $(a|b) =_x (c|d)$ implies only $a \vee b' = c \vee d'$, which does not imply equality of $(a|b)$ and $(c|d)$. Similarly for $x = tr$, because $(a|b) =_x (c|d)$ implies $ab = cd$, but not that $b = d$, which is required for equality. For $x = pm$, $(a|b) =_x (c|d)$ implies both $ab = cd$ and $a \vee b' = c \vee d'$. So $a'b = c'd$. Therefore $b = ab \vee a'b = cd \vee c'd = d$. So $(a|b) =_{pm} (c|d)$ implies (a|b) = (c|d). Each of the other two deductive relations $\leq_{m\wedge}$ and $\leq_{mv}$ are stronger than $\leq_{pm}$ and so imply equality too. For x=bo, (a|b) $=_x$ (c|d) trivially implies both b=d and ab=cd and so (a|b) = (c|d). For x=1, trivially (a|b) $=_x$ (c|d) means (a|b) = (c|d). Finally the remaining two deductive relations, $\leq_\wedge$ and $\leq_v$, are stronger than one of $\leq_{ap}$ or $\leq_{ip}$ and so (a|b) $=_x$ (c|d) implies that b = d. But (a|b) $=_v$ (c|d) implies ab = cd and thus that (a|b) = (c|d). Similarly, (a|b) $=_\wedge$ (c|d) implies a ∨ b' = c ∨ d'. Since b = d, (a ∨ b') b = (c ∨ d') d. That is, ab = cd, and so (a|b) = (c|d). That completes the proof of Theorem 6.2.18.

6.3 Construction of Deductively Closed Sets of Conditionals

Having described the hierarchy of deductive relations on conditionals it is now possible to define the set of implications of a set of conditionals with respect to such a deductive relation and to describe how to construct the deductively closed sets with respect to them.

6.3.1 Definition: Deductively Closed Set (DCS) of Conditionals

A subset H of conditionals is said to be a *deductively closed set* (DCS) *with respect to a deductive relation* $\leq_X$ if and only if H has both of the following properties:

1) If (a|b) ∈ H and (c|d) ∈ H   then (a|b) ∧ (c|d) ∈ H
2) If (a|b) ∈ H and (a|b) $\leq_X$ (c|d)   then (c|d) ∈ H

A *set of conditionals* having the first property is said to be *closed under conjunction,* or to have the conjunction property, and *a set of conditionals* having the second property is said to be *closed under deduction*, or to be *deductively closed*.

To show that all of these deductive relations have at least one closed set of conditionals the following are listed: The set H = $\mathcal{B}/\mathcal{B}$, the whole set of conditionals, although "improper" (inconsistent), is deductively closed with respect to any deductive relation $\leq_X$. H = {(q|p): p ≤ q} is deductively closed with respect to $\leq_{nf}$. H = {(1|1)} is deductively closed with respect to $\leq_v$. H(b) = {(x|y): y ≤ b} is deductively closed with respect to $\leq_{ip}$. H(b) = {(x|y): b ≤ y} is deductively closed with respect to $\leq_{ap}$. $H(b) = \{(x|y): b \leq xy\}$ is deductively closed with respect to $\leq_{tr}$.

The following theorem states that the intersection of two DCS's with respect to two different deductive relations is a DCS with respect to the deductive relation formed by combining the requirements of those two different deductive relations.

6.3.2 Intersection Theorem for DCS's

If H_x is a DCS of conditionals with respect to a deductive relation $\leq_x$ and H_y is a DCS of conditionals with respect to a deductive relation $\leq_y$, then $H_x \cap H_y$ is a deductively closed set of conditionals $H_{x \cap y}$ with respect to the *combined deductive relation* $\leq_{x \cap y}$ defined by:

$$(a|b) \leq_{x \cap y} (c|d) \quad \text{if \& only if} \quad (a|b) \leq_x (c|d) \quad \text{and} \quad (a|b) \leq_y (c|d)$$

That $\leq_{x \cap y}$ is a deductive relation is straightforward, and the proof of the theorem is also quite straightforward by passing from "and" in the meta-language to "$\cap$" in the object language. The converse is not true. Some DCS's with respect to a combined deductive relation are not the intersection of DCS's with respect to the component deductive relations.

In view of the hierarchical relationships between the various deductive relations presented here, the above theorem significantly simplifies matters since:

a. All non-trivial deductive relations in the hierarchy can be built up by combining the four elementary deductive relations.
b. The intersection of DCS's with respect to two deductive relations is a DCS with respect to the combined deductive relation.
c. For any initial subset J of conditionals, to determine the deductively closed sets generated by J with respect to the various combined deductive relations in the hierarchy, start with the intersections of the deductively closed sets with respect to the component deductive relations. All such intersections will be DCS's with respect to the combined deductive relation, but not all DCS's with respect to the combined deductive relation are intersections of DCS's with respect to their component deductive relations.

6.3.3 Conjunction of Conditionals Implies Disjunction
Note also that for any deductive relation $\leq_x$ with the Boolean Extension Property, $(a|b) \wedge (c|d) \leq_x (a|b) \vee (c|d)$ always holds. It then follows from either of the defining properties of a deductive relation that for any deductively closed set H

If $(a|b) \in H$ and $(c|d) \in H$ then $(a|b) \vee (c|d) \in H$

But it is not in general true (unless $\leq_v$ is also assumed) that

If $(a|b) \in H$ then for all $(c|d)$, $(a|b) \vee (c|d) \in H$

6.4 Generators of a Deductively Closed Set of Conditionals
In practice we are interested in determining what conditionals can be deduced, and in what sense deduced, from a given set J of conditionals.

6.4.1 Definition: Deductive Implications of a Set J of Conditionals
If J is any subset of conditionals, $H_x(J)$ will denote the smallest deductively closed set of conditionals with respect to $\leq_x$ that includes J. We say that $H_x(J)$ is the *deductive extension* of J with respect to $\leq_x$, or that J generates $H_x(J)$ with respect to $\leq_x$, or that J "$\leq_x$-implies" the DCS $H_x(J)$. A DCS is *principal* if it is generated by a single conditional.

6.4.2 Definition: Conjunction Property for Deductive Relations
A deductive relation $\leq_x$ has the *conjunction property* if and only if

$(a|b) \leq_x (c|d)$ and $(a|b) \leq_x (e|f)$ implies $(a|b) \leq_x (c|d) \wedge (e|f)$.

[Note: This is different from the property of conjunctive closure for a set of conditionals.]

6.4.3 Theorem: Principal Deductively Closed Sets
With respect to any deductive relation $\leq_x$ having the conjunction property the deductively closed set generated by a single conditional $(a|b)$ is the set of conditionals that subsume it with respect to the de-

ductive relation. That is, $H_x\{(a|b)\} = \{(y|z): (a|b) \leq_x (y|z)\}$. $H_x\{(a|b)\}$ will be denoted by $H_x(a|b)$.

Proof of Theorem 6.4.3 $H_x(a|b)$ is closed under conjunction. For suppose that $(c|d)$ and $(e|f)$ are in $H_x(a|b)$. So $(a|b) \leq_x (c|d)$ and $(a|b) \leq_x (e|f)$. Therefore $(a|b) \leq_x (c|d)(e|f)$, by the conjunction property of $\leq_x$. So $(c|d)(e|f) \in H_x(a|b)$. $H_x(a|b)$ obviously also has the deduction property by the transitivity of any deductive relation $\leq_x$. Therefore $H_x(a|b)$ is a DCS of conditionals. Clearly any DCS containing $(a|b)$ must also include $H_x(a|b)$. So $H_x(a|b)$ is the smallest DCS containing $(a|b)$.

6.4.4 Theorem: The Conjunction Property for $\leq_{ap}$, $\leq_{tr}$, $\leq_{nf}$, and $\leq_{ip}$

The four elementary deductive relations $\leq_{ap}$, $\leq_{tr}$, $\leq_{nf}$, and $\leq_{ip}$ on conditionals and their combinations, have the conjunction property of Definition 6.4.4.

Proof of Theorem 6.4.4 Suppose that $(a|b) \leq_{ap} (c|d)$ and $(a|b) \leq_{ap} (e|f)$. So $b \leq d$ and $b \leq f$. So $b \leq (d \wedge f) \leq (d \vee f)$. Therefore $(a|b) \leq_{ap} (c|d) \wedge (e|f) = (cdf' \vee d'ef \vee cdef \mid d \vee f)$ because $b \leq d \vee f$. Suppose next that $(a|b) \leq_{tr} (c|d)$ and $(a|b) \leq_{tr} (e|f)$. So $ab \leq cd$ and $ab \leq ef$. Therefore $(a|b) \leq_{tr} (c|d) \wedge (e|f)$ because $ab \leq (cd) \wedge (ef) \leq (cdf' \vee d'ef \vee cdef) \wedge (d \vee f)$. Suppose next that $(a|b) \leq_{nf} (c|d)$ and $(a|b) \leq_{nf} (e|f)$. So $(a \vee b') \leq (c \vee d')$ and $(a \vee b') \leq (e \vee f')$. Therefore $(a|b) \leq_{nf} (c|d) \wedge (e|f)$ because $(a \vee b') \leq (c \vee d') \wedge (e \vee f') = (cd \vee d') \wedge (ef \vee f') = (cdef \vee d'ef \vee cdf') \vee d'f'$, which is just $(cdf' \vee d'ef \vee cdef) \vee (d \vee f)'$. Fourthly, suppose that $(a|b) \leq_{ip} (c|d)$ and $(a|b) \leq_{ip} (e|f)$. So $b' \leq d'$ and $b' \leq f'$. Therefore $(a|b) \leq_{ip} (c|d) \wedge (e|f)$ because $b' \leq d' \wedge f' = (d \vee f)'$. Finally, Suppose that $(a|b) \leq_{x \cap y} (c|d)$ and $(a|b) \leq_{x \cap y} (e|f)$ where x and y are in $\{ap, tr, nf, ip\}$. So $(a|b) \leq_x (c|d)$ and $(a|b) \leq_y (c|d)$ and $(a|b) \leq_x (e|f)$ and $(a|b) \leq_y (e|f)$. Therefore $(a|b) \leq_x (c|d)$ and $(a|b) \leq_x (e|f)$ and so $(a|b) \leq_x (c|d) \wedge (e|f)$. Similarly $(a|b) \leq_y (c|d) \wedge (e|f)$. Therefore $(a|b) \leq_{x \cap y} (c|d) \wedge (e|f)$. That completes the proof of Theorem 6.4.4.

6.4.5 Theorem: The DCS Generated by One Conditional Proposition

If $\leq_x$ is one of the elementary deductive relations $\leq_{ap}, \leq_{tr}, \leq_{nf}$, and $\leq_{ip}$ or a deductive relation combining two or more of these, then the DCS generated by (a|b) with respect to $\leq_x$ is $H_x(a|b) = \{(y|z): (a|b) \leq_x (y|z)\}$.

Proof of Theorem 6.4.5 The proof follows immediately from Theorems 6.4.3 and 6.4.4.

With respect to each of the deductive relations in the hierarchy of Figure 6.1 it is now possible to completely describe the *principal* deductively closed sets of conditionals, that is, those generated by a single conditional (a|b).

6.4.6 The Principal DCS's with respect to $\leq_{ap}$, $\leq_{tr}$, $\leq_{nf}$, and $\leq_{ip}$

For deductive relation $\leq_{ap}$, the DCS generated by the single conditional (a|b) is $H_{ap}(a|b) = \{(c|d): (a|b) \leq_{ap} (c|d)\} = \{(c|d): b \leq d\}$, and so

$$H_{ap}(a|b) = \{(x \mid b \vee y): \text{any } x, y \text{ in } \mathcal{B}\}.$$

[Note that with respect to $\leq_{ap}$ by itself any conditional (a|b) generates (0|b) and also (0|d) for all d that include b. Thus $H_{ap}(a|b)$ is a partially "improper" (inconsistent) DCS because it includes all the zero conditionals $\{(0|y): b \leq y\}$.]

For deductive relation $\leq_{tr}$, the DCS generated by the single conditional (a|b) is

$$H_{tr}(a|b) = \{(c|d): (a|b) \leq_{tr} (c|d)\}$$
$$= \{(c|d): ab \leq cd\},$$

and so

$$H_{tr}(a|b) = \{(ab \vee x \mid ab \vee y): \text{any } x, y \text{ in } \mathcal{B}\}.$$

For deductive relation $\leq_{nf}$, the DCS generated by the single conditional (a|b) is

$$H_{nf}(a|b) = \{(c|d): (a|b) \leq_{nf} (c|d)\}$$
$$= \{(c|d): a \vee b' \leq c \vee d'\}$$
$$= \{(c|d): d(a \vee b') \leq c\}$$
$$= \{((d(a \vee b') \vee x) | d): \text{any } x, d \text{ in } \mathcal{B}\}$$
$$= \{((y(a \vee b') \vee x) | y): \text{any } x, y \text{ in } \mathcal{B}\},$$

and so

$$H_{nf}(a|b) = \{(ab \vee b' \vee x | y): \text{any } x, y \text{ in } \mathcal{B}\}.$$

For deductive relation $\leq_{ip}$, the DCS generated by the single conditional $(a|b)$ is $H_{ip}(a|b) = \{(c|d): (a|b) \leq_{ip} (c|d)\} = \{(c|d): b' \leq d'\} = \{(c|d): d \leq b\}$, and so

$$H_{ip}(a|b) = \{(x|by): \text{any } x, y \text{ in } \mathcal{B}\}.$$

[Note again that $H_{ip}(a|b)$ includes the conditional zeroes $(0|y)$ for all $y \leq b$, and so by itself generate only improper DCS's. These principal DCSs for the elementary deductive relations are given in non-reduced form but the reduced forms can easily be determined using $(a|b) = (ab | b)$.

The following result allows the principal DCS's of the deductive relations formed by combining two or more of the elementary deductive relations to be expressed as an intersection of principal DCS's of the elementary deductive relations. This result does not extend to DCS's generated by a set of conditionals.

6.4.7 Theorem: Principal DCS's for a Combined Deductive Relation

The principal DCS $H_{x \cap y}(a|b)$ generated by a single conditional $(a|b)$ with respect to a combination deductive relation $\leq_{x \cap y}$ is the intersection of the DCS's with respect to the component deductive relations $\leq_x$ and $\leq_y$. That is, $H_{x \cap y}(a|b) = H_x(a|b) \cap H_y(a|b)$. Thus every principal DCS of a combined deductive relation is the conjunction of DCS's with respect to the component deductive relations. (But some finite DCS's are not principal.)

Proof of Theorem 6.4.7 $H_{x \cap y}(a|b) = \{(c|d): (a|b) \leq_{x \cap y} (c|d)\} = \{(c|d): (a|b) \leq_x (c|d) \text{ and } (a|b) \leq_y (c|d)\} = H_x(a|b) \cap H_y(a|b)$.

6.4.8 Principal DCS's of Non-Elementary Deductive Relations

Using the Intersection Theorem 6.3.2 for DCS's with respect to two deductive relations, and Theorem 6.4.7 on principal DCS's, all the DCS's generated by a single conditional (a|b) by the non-elementary deductive relations can be determined as the intersection of the principal DCS's of the elementary deductive relation.

For deductive relation $\leq_v$, the DCS generated by the single conditional (a|b) is $H_v(a|b) = H_{ap \cap tr}(a|b) = H_{ap}(a|b) \cap H_{tr}(a|b)$. So by combining constraints expressed above for $H_{ap}(a|b)$ and $H_{tr}(a|b)$,

$$H_v(a|b) = \{(ab \vee x \mid b \vee y): \text{any } x, y \text{ in } \mathcal{B}\}.$$

For deductive relation $\leq_{pm}$, the DCS generated by the single conditional (a|b) is $H_{pm}(a|b) = H_{tr \cap nf}(a|b) = H_{tr}(a|b) \cap H_{nf}(a|b)$. So by combining constraints expressed above for $H_{tr}(a|b)$ and $H_{nf}(a|b)$,

$$H_{pm}(a|b) = \{(ab \vee b' \vee x \mid ab \vee y): \text{any } x, y \text{ in } \mathcal{B}\}.$$

For deductive relation $\leq_\wedge$, the DCS generated by the single conditional (a|b) is $H_\wedge(a|b) = H_{nf \cap ip}(a|b) = H_{nf}(a|b) \cap H_{ip}(a|b)$. So by combining constraints expressed above for $H_{nf}(a|b)$ and $H_{ip}(a|b)$, $H_\wedge(a|b) = \{((ab \vee b' \vee x) \mid by): \text{any } x, y \text{ in } \mathcal{B}\}$, which can be simplified as

$$H_\wedge(a|b) = \{(ab \vee x \mid by): \text{any } x, y \text{ in } \mathcal{B}\}.$$

For deductive relation $\leq_{ec}$, the DCS generated by the single conditional (a|b) is simply $\{(x|y): y = b\}$. So

$$H_{ec}(a|b) = \{(x|b): \text{any } x \text{ in } \mathcal{B}\}.$$

For deductive relation $\leq_{mv}$, the DCS generated by the single conditional (a|b) is $H_{mv}(a|b) = H_{ap \cap pm}(a|b) = H_{ap}(a|b) \cap H_{pm}(a|b)$. So by combining constraints expressed above for $H_{ap}(a|b)$ and $H_{pm}(a|b)$,

$$H_{mv}(a|b) = \{((ab \vee b' \vee x) \mid (b \vee y)): \text{any } x, y \text{ in } \mathcal{B}\}.$$

(Note that $H_{mv}(a|b)$ can also be expressed as $H_{ap \cap nf}(a|b)$ or as $H_{v \cap nf}(a|b)$ or as $H_{v \cap pm}(a|b)$ with equivalent results.)

For deductive relation $\leq_{m \wedge}$, the DCS generated by the single conditional (a|b) is $H_{m \wedge}(a|b) = H_{tr \cap ip}(a|b) = H_{tr}(a|b) \cap H_{ip}(a|b)$. So by combining constraints expressed above for $H_{tr}(a|b)$ and $H_{ip}(a|b)$, $H_{m \wedge}(a|b) = \{((ab \vee x) \mid (ab \vee yb)): \text{any } x, y \text{ in } \mathcal{B}\}$. So,

$$H_{m \wedge}(a|b) = \{(a \vee x \mid ab \vee yb): \text{any } x, y \text{ in } \mathcal{B}\}.$$

These are all the conditionals whose condition is an event between ab and b, and whose conclusion includes a. (Note again that $H_{m \wedge}(a|b)$ can also be expressed as $H_{tr \cap \wedge}(a|b)$ or as $H_{pm \cap ip}(a|b)$ or as $H_{pm \cap \wedge}(a|b)$ with equivalent results.)

For the deductive relation $\leq_{bo}$ the DCS generated by the single conditional (a|b) is

$$H_{bo}(a|b) = \{((a \vee x) \mid b): \text{any } x \text{ in } \mathcal{B}\}.$$

Having described all the principal DCS's of the 11 non-trivial deduction relations in the hierarchy, they can be used to build up the DCS's generated by more than a single conditional. In this regard Dubois and Prade [96, Definition 1, p1719] adopt a theorem of Adams [2, Theorem 1, p52] expressed in probability context in order to define the logical entailment of a set of conditionals J. There construct is similar to what below is called the conjunctive closure.

6.4.9 Definition: Conjunctive Closure of a Set of Conditionals

If J is a set of conditionals then the *conjunctive closure* C(J) of J is the set of all conjunctions of any finite subset of J. In symbols,

$$C(J) = \left\{ \bigwedge_{i=1}^{n} (a|b)_i : n \text{ finite}, (a|b)_i \in J \right\}$$

6.4.10 Deductive Extension Theorem

For all the elementary deductive relations $\leq_X$ and their combinations, except for $\leq_{tr}$ and $\leq_v$, the deductively closed set $H_X(J)$ with respect to $\leq_X$ of a set J of conditionals is the set of all conditionals implied with respect to $\leq_X$ by some member of the conjunctive closure C(J) of J. That is,

$$H_X(J) = \{(c|d): (a|b) \leq_X (c|d), (a|b) \in C(J)\}$$

That is,

$$H_X(J) = \{(c|d): \exists\, (a|b)_i \in J, i = 1 \text{ to } n \text{ (finite) such that } \bigwedge_{i=1}^{n}(a|b)_i \leq_X (c|d)\}$$

Proof of Theorem 6.4.10 First property 2) of Definition 6.3.1: If (a|b) $\in H_X(J)$ and (a|b) $\leq_X$ (c|d), then there exists (q|p) $\in$ C(J) such that (q|p) $\leq_X$ (a|b). By transitivity of $\leq_X$, (q|p) $\leq_X$ (c|d). Therefore (c|d) $\in H_X(J)$. That shows property 2) for any deductive relation $\leq_X$. To show property 1), suppose that (a|b) $\in H_X(J)$ and (c|d) $\in H_X(J)$. So there exist (q|p) $\in$ C(J) and (s|r) $\in$ C(J) such that (q|p) $\leq_X$ (a|b) and (s|r) $\leq_X$ (c|d). Now, (q|p) ∧ (s|r) $\in$ C(J) because the conjunction of two finite conjunctions of elements of J is a finite conjunction of elements of J. It follows (by the next lemma) that except for x $\in$ {tr, v}, (q|p) ∧ (s|r) $\leq_X$ (a|b) ∧ (c|d). Therefore (a|b) ∧ (c|d) $\in H_X(J)$. So property 1) holds. Finally, $H_X(J)$ is the smallest deductively closed subset with respect to $\leq_X$ that includes J because $H_X(J)$ includes C(J), and C(J) includes J. So $H_X(J)$ includes J. Secondly, any deductively closed set that includes J must, by repeated application of property 1), include all finite conjunctions of elements of J, and so, by property 2), must include any (c|d) for which there is an (a|b) $\in$ J with (a|b) $\leq_X$ (c|d). Concerning the deductive relations $\leq_X$, for x $\in$ {tr, v}, a counter example to the theorem is provided by the set J = {(1|b), (1|b')} in the nine element conditional algebra $\mathcal{B}/\mathcal{B}$ generated by Boolean algebra $\mathcal{B}$ = {1, b, b', 0}. (See the four DCS's in Table 6.1.) For this set J, C(J) = {(1|b), (1|b'), 1}, and so by the theorem $H_{tr}(J)$ would be {(e|f): (a|b) $\leq_{tr}$ (e|f), (a|b) $\in$ C(J)} = {1, (1|b), (1| b'), b, b'}. However, this latter set is not a DCS with respect to $\leq_{tr}$ because it obviously generates 0 from b and b' and so is not yet conjunctively closed. The same counter example also works for $\leq_v$. See Table 6.3, DCS's #4, #5 and #11. Here again, as for $\leq_{tr}$, the union of the principal DCS's generated by the conditionals in C(J) = {(1|b), (1| b'), 1} is not a DCS.

6.4.11 Lemma for Deductive Extension Theorem
Let $\leq_x$ be any deductive relation in the hierarchy other than $\leq_{tr}$ or $\leq_v$. If (q|p) $\leq_x$ (a|b) and (s|r) $\leq_x$ (c|d), then (q|p) ∧ (s|r) $\leq_x$ (a|b) ∧ (c|d). For x $\in$ {tr, v}, this is not necessarily true.

Proof of Lemma 6.4.11 First a counter-example for x $\in$ {tr, v}: Consider that (0|0) $\leq_v$ (0|1) and (1|1) $\leq_v$ (1|1). Now (0|0) ∧ (1|1) = (1|1) and (0|1) ∧ (1|1) = (0|1). But (1|1) $\leq_{tr}$ (0|1) is always false and so too is the stronger (1|1) $\leq_v$ (0|1). Concerning x=ap, suppose (q|p) $\leq_{ap}$ (a|b) and (s|r) $\leq_{ap}$ (c|d). So p $\leq$ b and r $\leq$ d, and so p ∨ r $\leq$ b ∨ d. Therefore

$$(q|p) \wedge (s|r) = ((qpr' \vee p'sr \vee qpsr) | (p \vee r))$$
$$\leq_{ap} ((abd' \vee b'cd \vee abcd) | (b \vee d)) = (a|b) \wedge (c|d).$$

A similar argument with the inequalities reversed works for x = ip. Concerning x=nf, suppose that both (q|p) $\leq_{nf}$ (a|b) and (s|r) $\leq_{nf}$ (c|d). Then
$$(q|p) \wedge (s|r) = (qpr' \vee p'sr \vee qpsr) | (p \vee r)$$
$$\leq_{nf} (abd' \vee b'cd \vee abcd) | (b \vee d) = (a|b) \wedge (c|d)$$
since
$$(qpr' \vee p'sr \vee qpsr) \vee (p \vee r)'$$
$$= (qp \vee p')(sr \vee r')$$
$$= (q \vee p')(s \vee r') \leq (a \vee b')(c \vee d')$$
$$= (abd' \vee b'cd \vee abcd) \vee (b \vee d)'.$$

That shows the result for x = nf. Since the result holds for x = nf and for x = ip, the result holds for x = ∧, which just combines these two elementary deductive relations. Now suppose the hypothesis holds for x = pm. So the four inequalities (q ∨ p′) ≤ (a ∨ b′), qp ≤ ab, (s ∨ r′) ≤ (c ∨ d′) and sr ≤ cd hold. Again, the result holds for x = nf. Therefore, by the first part of the proof, the result holds for x = pm providing (qp)(s ∨ r′) ∨ (q ∨ p′)(sr) ≤ (ab)(c ∨ d′) ∨ (cd)(a ∨ b′), and the latter easily follows from these four inequalities. The deductive relation ≤$_{mv}$ is made up of the two elementary deductive relations ≤$_{ap}$ and ≤$_{nf}$, for which the result holds. Therefore the result holds for ≤$_{mv}$. The deductive relation ≤$_{m∧}$ is just the conjunction of the two deductive relations ≤$_{pm}$ and ≤$_∧$ for which the result holds. So it holds too for ≤$_{m∧}$. Similar statements can be made for x = ec and bo, or proved directly. That completes the proof of the lemma.

It may at first seem disappointing that it is necessary to take every possible finite conjunction of the members of a given generating set J of conditionals in order to determine the deductive consequences of J with respect to most deductive relations. Deduction monotonicity requires that additional conditional information will have additional deductive implications. The different *premises* of the members of J need to be disjoined (∨) in all possible ways, as they are when conditionals are conjoined or disjoined according to the operations. A single conditional, say the conjunction of all members of a finite set J, will have the largest possible premise, namely the disjunction of all the premises of the members of J. However, as proved in the next theorem, for three of the deductive relations this single conditional by itself generates the same DCS as J.

6.4.12 Corollary of Deductive Extension Theorem
Since $H_x(J)$ is the union of all conditionals that are implied by some individual member of C(J), it follows by Theorem 6.4.3 on principal DCS's that for all deductive relations ≤$_x$ in the hierarchy except for ≤$_{tr}$ and ≤$_v$, the DCS generated by a set J of conditionals is the union of the principal DCS's generated by the individual members of the set D(J) of all finite conjunctions of the conditionals in J. Except for x ∈ {tr, v}

$$H_x(J) = \bigcup_{(a|b)\, \in\, C(J)} H_x(a|b)$$

That is, except for $x \in \{tr, v\}$ the deductively closed set with respect to $\leq_x$, generated by a subset J of conditionals is the set of all conditionals implied with respect to $\leq_x$ by some member of the conjunctive closure $C(J)$ of J.

For most deductive relations $\leq_x$ it is necessary in general to first determine the conjunctive closure $C(J)$ of a finite set of conditionals J in order to determine the DCS $H_x(J)$ of J. However for the non-falsity, inapplicability and conjunctive deductive relations, that is for $x \in \{nf, ip, \wedge\}$, the DCS of J is $H_x(J) = H_x(a|b)$, where (a|b) is the single conditional formed by conjoining all the conditionals in J.

6.4.13 Theorem: All finite DCS's of $\leq_{nf}$, $\leq_{ip}$, and $\leq_\wedge$ are Principal

With respect to the three deductive relations $\leq_{nf}$, $\leq_{ip}$, and $\leq_\wedge$ the DCS of a finite set of conditionals J is principal and is generated by the single conditional formed by conjoining all the conditionals in J. That is, for $x \in \{nf, ip, \wedge\}$, the DCS with respect to $\leq_x$ generated by a finite set J of conditionals is principal, and is generated by the conditional proposition $(A|B)_J$ defined by:

$$A|B)_J = \bigwedge_{(a|b)_i \in J} (a|b)_i$$

This is not true for the other deductive relations in the hierarchy, and in fact their finite DCS's are not in general principal.

Proof of Theorem 6.4.13 Let $x \in \{nf, ip, \wedge\}$, and suppose (A|B) is the conjunction of all the conditionals in the set J of conditionals. Then $H_x(J) = H_x(A|B)$ because with respect to $\leq_x$, $(A|B) \leq_x (c|d)$ for all (c|d) in $C(J)$. Briefly, this follows from the fact that for these 3 deductive relations the conjunction of two conditionals always implies each of the component conditionals.

In detail, since J is a finite set of conditionals $\{(a|b)_i\}$, it follows by repeated application of property 1) of Definition 6.3.1 that for all deductive relations $\leq_x$, $(A|B)_J$ is a conditional proposition in the DCS with respect to $\leq_x$ generated by J. Furthermore, for x in $\{nf, ip, \wedge\}$, $(A|B)_J \leq_x (a|b)_i$, for all $(a|b)_i \in J$.

For x = nf, this last assertion follows from the fact that $(a|b) \wedge (c|d) = [(a \vee b')(c \vee d') | (b \vee d)]$, which is, by definition, not false on $(a \vee b')(c \vee d') \vee (b \vee d)' = (a \vee b')(c \vee d')$, and $(a \vee b')(c \vee d') \leq (a \vee b')$. So $(a|b) \wedge (c|d) \leq_{nf} (a|b)$ and $(a|b) \wedge (c|d) \leq_{nf} (c|d)$. Repeated application then yields the result for x = nf.

For x = ip, this follows because conjunction of conditionals always yields a resulting conditional whose condition includes the conditions of all the components of the conjunction, and therefore the conjunction will imply each of the component conditionals with respect to x = ip.

For x = $\wedge$, by definition, the conjunction of conditionals implies with respect to $\leq_\wedge$ the components of the conjunction. Alternately, since the defining properties of $\leq_\wedge$ consist of the combined characteristics of $\leq_{ip}$ and $\leq_{nf}$, it follows that $\leq_\wedge$ too has the property.]

Summarizing, for x = ip, nf or $\wedge$, the principal DCS with respect to $\leq_x$ generated by $(A|B)_J$ includes all the conditional propositions of J. So with respect to $\leq_x$, $(A|B)_J$ generates all the conditional propositions of J and is generated by J. That is, $H_x(J) = H_x(A|B)_J$. For the other deductive relations in the hierarchy, Section 4.3 will provide examples of non-principal DCS's, which therefore cannot be generated by any single conditional including $(A|B)_J$. That completes the proof of Theorem 6.4.13.

When a new conditional is adjoined to a collection of conditionals, or if two sets of conditionals are combined, the resulting collection has new deductions.

6.4.14 Theorem on Additional Deductive Information

For all deductive relations $\leq_x$ in the hierarchy except for $x \in \{tr, v\}$, the deductive extension of a DCS J and an additional conditional proposition (c|d) is

$$H_x(J \cup \{(c|d)\}) = \{(q|p): \exists\, (a|b) \in J \text{ for which } (a|b) \wedge (c|d) \leq_x (q|p)\}$$

and more generally, if K is another DCS of conditional propositions with respect to $\leq_x$, $x \notin \{tr, v\}$ and in the hierarchy, then

$$H_x(J \cup K) = \{(q|p): \exists\, (a|b) \in J \text{ and } (c|d) \in K \text{ such that } (a|b) \wedge (c|d) \leq_x (q|p)\}$$

Which is the set of conditionals implied with respect to $\leq_x$ by the conjunction of some conditional in J with some conditional in K.

Proof of Theorem 6.4.14 $H_x(J \cup K)$ is a DCS since if $(q|p) \in H_x(J \cup K)$ and $(s|r) \in H_x(J \cup K)$ then there are $(a|b) \in J$ and $(c|d) \in K$ with $(a|b) \wedge (c|d) \leq_x (q|p)$. Similarly there exist $(e|f) \in J$ and $(g|h) \in K$ with $(e|f) \wedge (g|h) \leq_x (s|r)$. So by the Lemma of the Extension Theorem $(a|b) \wedge (c|d) \wedge (e|f) \wedge (g|h) \leq_x (q|p) \wedge (s|r)$. By the commutative law for conditionals this can be expressed as $(a|b) \wedge (e|f) \wedge (c|d) \wedge (g|h) \leq_x (q|p) \wedge (s|r)$, with $(a|b) \wedge (e|f) \in J$ and $(c|d) \wedge (g|h) \in K$. So $(q|p) \wedge (s|r) \in H_x(J \cup K)$. That shows that $H_x(J \cup K)$ is closed under conjunction. Now suppose $(q|p) \in H_x(J \cup K)$ and that $(q|p) \leq_x (s|r)$. So there are $(a|b) \in J$ and $(c|d) \in K$ with $(a|b) \wedge (c|d) \leq_x (q|p)$. By transitivity it easily follows that $(a|b) \wedge (c|d) \leq_x (s|r)$. So $(s|r) \in H_x(J \cup K)$. That completes the proof.

6.4.15 Definition: The Conjunction (J ∧ K) of Sets J and K

By defining as usual $(J \wedge K)$ to be $\{(a|b) \wedge (c|d): (a|b) \in J, (c|d) \in K\}$, it follows that for all deductive relations $\leq_x$ in the hierarchy except $x \in \{tr, v\}$, $(J \cup K) \subseteq (J \wedge K) \subseteq H_x(J \cup K)$, but the equalities may not hold. The deductively closed set generated by the union of two DCS's can be something more than the simple conjunction of the conditionals of one DCS with those of the other DCS. This simple conjunction may not be a DCS. Here again, there is a difference between the situation for conditional propositions and the situation for Boolean propositions. In the Boolean case, $J \wedge K = H_x(J \cup K)$ always holds. That is, for Boolean ideals J and K, $J \wedge K = (J \cup K)$, the ideal generated by the union of J and K. But even in the Boolean case the conjunction $J \wedge K$ of two DCS's (ideals) can be larger than the simple union of the component ideals.

6.5 The Exceptional Deductive Relations $\leq_{tr}$ and $\leq_v$

Having described the DCS's of all but two of the deductive relations in the hierarchy, it remains to solve for the DCS's with respect to the exceptional deductive relations $\leq_{tr}$ and $\leq_v$. See Calabrese [71].

6.5.1 Theorem: The DCS's of the Deductive Relation $\leq_{tr}$

For the deductive relation $\leq_{tr}$, the DCS generated by two conditionals (a|b) and (c|d) is principal and generated by the conjunction abcd of the components of the conditionals. That is,

$$H_{tr}\{(a|b), (c|d)\} = H_{tr}(abcd)$$

Proof of Theorem 6.5.1 Let $H = H_{tr}\{(a|b), (c|d)\}$. Note that $(a|b) =_{tr}$ ab and $(c|d) =_{tr}$ cd. Since $(a|b) \leq_{tr}$ ab therefore ab $\in$ H. Since $(c|d) \leq_{tr}$ cd therefore cd $\in$ H. Since both ab and cd are in H therefore $(ab)(cd) \in H$. So $H_{tr}(abcd) \subseteq H$. Conversely, since abcd $\leq_{tr}$ ab and ab $\leq_{tr}$ (a|b) therefore (a|b) $\in H_{tr}(abcd)$. Similarly, (c|d) $\in H_{tr}(abcd)$. Since both (a|b) and (c|d) are in $H_{tr}(abcd)$ therefore $H = H_{tr}\{(a|b), (c|d)\} \subseteq H_{tr}(abcd)$. So $H = H_{tr}(abcd)$.

6.5.2 Corollary:

For $\leq_{tr}$ and all a & c, the DCS of {a, c} is the DCS of $\{a \wedge c\}$. That is, $H_{tr}\{a, c\} = H_{tr}(ac)$. $H_{tr}\{ab, cd\} = H_{tr}(abcd)$.

Proof of Corollary 6.5.2 Since ac $\leq$ a and ac $\leq$ c, both a and c are in $H_{tr}(ac)$. So $H_{tr}\{a,c\} \subseteq H_{tr}(ac)$. Conversely, ac $\in H_{tr}\{a,c\}$. So $H_{tr}\{a,c\} = H_{tr}(ac)$. Replacing a with ab and c with cd yields that $H_{tr}\{ab, cd\} = H_{tr}(abcd)$.

6.5.3 Corollary: For $\leq_{tr}$ the DCS of a Finite Number of Conditionals

If $J = \{(a_i \mid b_i): i = 1, 2 \ldots, n\}$ then

$$H_{tr}(J) = H_{tr}(\prod_{i=1}^{n} a_i b_i) = H_{tr}\{a_1 b_1, a_2 b_2, \ldots a_n b_n\}.$$

Proof of Corollary 6.5.3 The same argument works as in Theorem 6.5.1 and Corollary 6.5.2.

Note that $H_{tr}(ab) \wedge H_{tr}(cd) \subseteq H_{tr}(abcd)$ but in general $H_{tr}(ab) \wedge H_{tr}(cd) \neq H_{tr}(abcd)$ since conjunctions don't necessarily imply their conjuncts.

Note that while $\{ab, cd\}$ and $(abcd)$ both generate $H_{tr}\{(a|b), (c|d)\}$, the conjunction $(a|b) \wedge (c|d)$ may not. In general, $H_{tr}(a|b) \wedge (c|d)) \subseteq H_{tr}\{(a|b), (c|d)\}$.

6.5.4 Corollary: For $\leq_{tr}$, the DCS's of $(a|b) \wedge (c|d)$ and of $\{(a|b),(c|d)\}$

$H_{tr}((a|b)(c|d)) = H_{tr}\{(a|b), (c|d)\}$ if and only if $ab \leq d$ and $cd \leq b$. That is, if the truth of one conditional implies the applicability of the other conditional.

Proof of Corollary 6.5.4 $H_{tr}((a|b)(c|d)\} = H_{tr}(abd' \vee b'cd \vee abcd \mid b \vee d) = H_{tr}(abd' \vee b'cd \vee abcd)$. Since $abcd \leq abd' \vee b'cd \vee abcd$, $H_{tr}(abd' \vee b'cd \vee abcd \mid b \vee d) \subseteq H_{tr}(abcd)$ with equality if $abd' = 0 = b'cd$, that is, if $ab \leq d$ and $cd \leq B$. but if either $abd' \neq 0$ or $b'cd \neq 0$ then $H_{tr}(abd' \vee b'cd \vee abcd \mid b \vee d)$ does not contain $abcd$ since all members of $H_{tr}(abd' \vee b'cd \vee abcd \mid b \vee d)$ have both components above $(abd' \vee b'cd \vee abcd)$ and $(abd' \vee b'cd \vee abcd) > abcd$.

For completeness there is also the following theorem concerning the DCS's with respect to the deductive relation $\leq_{ap}$.

6.5.5 Theorem: The Union of Principal DCS for Deductive Relation $\leq_{tr}$

For the deductive relation $\leq_{ap}$, the DCS generated by two conditionals (a|b) and (c|d) is $H_{ap}\{(a|b), (c|d)\} = H_{ap}(a|b) \cup H_{ap}(c|d)$. In general, if $J = \{(a_i|b_i): i = 1, 2 \ldots n\}$ then $H_{ap}(J) = H_{ap}(a_1|b_1) \cup H_{ap}(a_2|b_2) \cup \ldots \cup H_{ap}(a_n|b_n)$.

Proof of Theorem 6.5.5 By the Deductive Extension Theorem, or directly, $H_{ap}\{(a|b), (c|d)\} = H_{ap}(a|b) \cup H_{ap}(c|d) \cup H_{ap}((a|b)(c|d))$. But (a|b) $\leq_{ap}$ (a|b)(c|d) because $b \leq b \vee d$. Therefore $H_{ap}((a|b)(c|d)) \subseteq H_{ap}(a|b)$. So $H_{ap}\{(a|b), (c|d)\} = H_{ap}(a|b) \cup H_{ap}(c|d)$. In general, all conjunctions of members of J have antecedents that are implied by the antecedent of any one of the conjuncts. So all conjunctions of members of J are in the union $H_{ap}(a_1|b_1) \cup H_{ap}(a_2|b_2) \cup \ldots \cup H_{ap}(a_n|b_n)$, and this union is easily a DCS.

6.5.6 Theorem:

For the deductive relation $\leq_v$ the DCS generated by two conditionals (a|b) and (c|d) is $H_v\{(a|b), (c|d)\} =$

$$H_v(a|b) \cup H_v(c|d) \cup H_v(abcd \mid b \vee d).$$

Proof of Theorem 6.5.6 Let H denote the right hand side of the above equation. It will be shown that H is the smallest DCS containing both (a|b) and (c|d). Clearly H contains both (a|b) and (c|d) since it includes $H_v(a|b)$ and $H_v(c|d)$. Concerning the deduction property, suppose that (w|z) $\in$ H and (w|z) $\leq_v$ (s|t). Then (w|z) is in $H_v(a|b)$ or $H_v(c|d)$ or $H_v(abcd \mid b \vee d)$, and by the deduction property for these DCS's, so too is (s|t) in that DCS. So (s|t) $\in$ H. Concerning the conjunction property, suppose (e|f) and (g|h) are both in h. To show that their conjunction must also be in h, consider cases. If both are in any one of $H_v(a|b)$, $H_v(c|d)$ or $H_v(abcd \mid b \vee d)$, then their conjunction (e|f)(g|h) will also be in that one since they are each DCS's. So in those cases (e|f)(g|h) $\in$ H. If instead, (e|f) $\in$ $H_v(a|b)$ and (g|h) $\in$ $H_v(c|d)$, then (e|f) = (ab $\vee$ y_1 | b $\vee$ z_1) and (g|h) = (cd $\vee$ y_2 | d $\vee$ z_2) for some events y_1, z_1, y_2, z_2. (e|f) reduces to (ab $\vee$ y_1(b $\vee$ z_1) | b $\vee$ z_1) and (g|h) reduces to (cd $\vee$ y_2(d $\vee$ z_2) | d $\vee$ z_2). So

$(e|f)(g|h) = [(ab \vee y_1 (b \vee z_1))(d \vee z_2)' \vee (cd \vee y_2 (d \vee z_2))(b \vee z_1)'$
$\qquad \vee \quad (ab \vee y_1(b \vee z_1))(cd \vee y_2(d \vee z_2)) \mid b \vee d \vee z_1 \vee z_2 \,]$
$= [(ab)(cd) \vee \ldots \mid b \vee d \vee z_1 \vee z_2 \,] \in H_V(abcd \mid b \vee d) \subseteq H.$

So again (e|f)(g|h) is in H. If instead (e|f) ∈ H_V(a|b) and (g|h) ∈ H_V(abcd | b ∨ d) then a similar computation yields that (e|f)(g|h) ∈ H_V(abcd | b ∨ d) ⊆ H. By symmetry, the same result holds in case (e|f) ∈ H_V(c|d) and (g|h) ∈ H_V(abcd | b ∨ d). So in all cases (e|f)(g|h) ∈ H. Therefore H is a DCS. Finally, H is the smallest DCS containing (a|b) and (c|d) because any DCS containing them must contain (ab | b ∨ d) since (a|b) $\leq_V$ (ab | b ∨ d). Similarly (cd | b ∨ d) is in any DCS containing (a|b) and (c|d). Therefore (ab | b ∨ d)(cd | b ∨ d) = (abcd | b ∨ d) must be in any DCS containing (a|b) and (c|d). Therefore H = H_V(a|b) ∪ H_V(c|d) ∪ H_V(abcd | b ∨ d) must be the smallest DCS with respect to $\leq_V$ containing both (a|b) and (c|d). That completes the proof of Theorem 6.5.6.

6.5.7 Corollary:
For three conditionals (a|b), (c|d) and (e|f) the DCS generated with respect to the deductive relation $\leq_V$ is $H_V\{(a|b), (c|d), (e|f)\} =$

$H_V(a|b) \cup H_V(c|d) \cup H_V(e|f)$
$\cup\, H_V(abcd \mid b \vee d) \cup H_V(abef \mid b \vee f) \cup H_V(cdef \mid d \vee f)$
$\cup\, H_V(abcdef \mid b \vee d \vee f).$

Proof of Corollary 6.5.7 Let H denote the right hand side of the above equation. Clearly H contains the three conditionals (a|b), (c|d) and (e|f). H is also a DCS since it has the deduction and conjunction properties. It has the deduction property since if (g|h) ∈ H and (g|h) $\leq_V$ (r|s) then (g|h) is at least one of the seven DCS's whose union is H. So (r|s) is in at least one of the seven DCS's by the deduction property applied to that DCS. So H has the deduction property. Concerning the conjunction property, suppose (g|h) and (r|s) are in H. Then by considering cases similar to those in the proof of the theorem it follows that (g|h)(e|f) is also in H. It is also clear from the theorem that any DCS containing the three conditionals must also contain the conjunctions (a|b)(c|d), (c|d)(e|f), and (c|d)(e|f) and so must contain (abcd | b ∨ d), (abef | b ∨ f) and (cdef | d ∨ f) as shown in the proof of the theorem. Furthermore, if a DCS contains these latter three conditionals, then it must contain the conjunction of any two of them, which is (ab-cdef | b ∨ d ∨ f). So H is the smallest DCS that contains the original three conditionals.

6.6 Non-Elementary Examples of Deductively Closed Sets
With Theorem 6.4.14 it is easy to specify many non-elementary examples of DCS's with respect to various deductive relations.

6.6.1 Example on Transitivity
Consider the set J consisting of two uncertain conditionals (a|b) and (b|c). Then the conjunctive closure $C(J) = \{(a|b), (b|c), (a|b)(b|c)\} = \{(a|b), (b|c), (ab | b \vee c)\}$. For $x \in \{nf, ip, \wedge\}$, by Theorem 6.4.7 on principal DCS's, the DCS generated by J is $H_x(J) = H_x(ab | b \vee c)$. So using the results of Sections 6.4.6 – 6.4.8,

$$H_{ip}(J) = \{(y | (b \vee c)z): \text{any } y, z \text{ in } \mathcal{B}\}$$
$$H_{nf}(J) = \{(ab \vee b'c' \vee y | z): \text{any } y, z \text{ in } \mathcal{B}\}$$
$$H_\wedge(J) = \{(ab \vee y | (b \vee c)z): \text{any } y, z \text{ in } \mathcal{B}\}$$

Notice that $(a|c) \in H_{nf}(J)$ by setting $y = ab'$ and $z = c$. In that case $(ab \vee b'c' \vee y | z) = (ab \vee ab' \vee b'c' | c) = (a \vee b'c' | c) = (a|c)$. Thus with respect to the non-falsity deductive relation $\leq_{nf}$, the conditional (a|c), as expected, is implied by (a|b) and (b|c). When (a|b) and (b|c) are non-false then so is (a|c). $H_{nf}(J)$ is the set of all conditionals whose conclusion includes the truth of (a|b) and also the conjoint inapplicability of both (a|b) and (b|c). By similar arguments (a|c) is in $H_{ip}(J)$ and also in $H_\wedge(J)$.

For the elementary deductive relations $\leq_x$ or some combination of them except for $\leq_{tr}$ and $\leq_v$, by Corollary 6.4.12 the DCS generated by J is $H_x(J) = H_x(a|b) \cup H_x(b|c) \cup H_x(ab | b \vee c)$.

Now let x = pm. That is, consider the deductions of J with respect to the probabilistically monotonic deductive relation $\leq_{pm}$. Since (ab | b ∨ c) $\leq_{pm}$ (a|b), therefore $H_{pm}(ab | b \vee c) \supseteq H_{pm}(a|b)$. Thus, $H_{pm}(J) = H_{pm}(b|c) \cup H_{pm}(ab | b \vee c)$. So by results of Section 6.4.8, $H_{pm}(J) = \{bc \vee c' \vee y | bc \vee z): \text{any } y, z \text{ in } b\} \cup \{ab \vee b'c' \vee z | ab \vee z): \text{any } y, z \text{ in } b\}$. Note that (a|c) is not necessarily a member of $H_{pm}(J)$.

Furthermore, since $\leq_{pm}$ is probabilistically monotonic, all the conditionals in $H_{pm}(b|c) = \{bc \vee c' \vee y \mid bc \vee z)$: any y, z in b} have conditional probability no less than P(b|c), and all the conditionals in $H_{pm}(ab \mid b \vee c) = \{ab \vee b'c' \vee z \mid ab \vee z)$: any y, z in b} have conditional probability no less than P(ab | b ∨ c).

6.6.2 Applications of the Deductive Relations

We are used to making Boolean deductions in essentially one way. We show "A implies B" by showing that event A is a subset of event B, that every instance of A is an instance of B. But conditionals have two components, which complicates things, and results in several different kinds of implication, each good for a different purpose depending upon what properties one wants to imply in a deduced conditional. It may be that one wishes to imply the simple truth of one conditional (c|d) from another one (a|b), in which case, the deductive relation $\leq_{tr}$ would be appropriate. One could then conclude that P(ab) ≤ P(cd). If on the other hand one wishes to deduce the non-falsity of one conditional from another, then $\leq_{nf}$ would be the appropriate deductive relation. Every instance of (a|b) being true or inapplicable is an instance of (c|d) being true or inapplicable, and no matter what, P(a ∨ b') ≤ P(c ∨ d') for the two conditionals. This implication would be appropriate if one wished just to preserve Boolean non-falsily for logical purposes but was not concerned about the conditional probability of the conditionals when they were partially false. The implication $\leq_{pm}$ combines the defining characteristics of $\leq_{tr}$ and $\leq_{nf}$ so that both characteristics must be true for the implication to hold, in which case it follows that P(a|b) ≤ P(c|d). If instead of conditional probability, one wishes to have that (a|b) ∧ (c|d) = (a|b) whenever (a|b) implies (c|d)), as is always true in Boolean algebra, then one would use the deductive relation $\leq_\wedge$. Then one could only say that P(a ∨ b') ≤ P(c ∨ d') and that P(b') ≤ P(d'). On the other hand, if one wishes to have the property (a|b) ∨ (c|d) = (c|d) whenever (a|b) implies (c|d), as is always true in (unconditioned) Boolean algebra, then $\leq_v$ would be ap-propriate combining the characteristic of $\leq_{tr}$ with that of $\leq_{ap}$ and then it would also follow that P(ab) ≤ P(cd) and P(b) ≤ P(d). At the expense of another requirement for the deduction of (c|d) by (a|b), one can have the advantages of $\leq_{pm}$ combined with those of $\leq_\wedge$ in $\leq_{m\wedge}$, or the advantages of $\leq_{pm}$ combined with those of $\leq_v$ in $\leq_{mv}$. Boolean deduction ≤ extended to conditionals $\le_{bo}$ (see Figure 6.1) combines the characteristic properties of all four of the elementary deductive relations $\le_x$, x in {ap, tr, nf, ip}. It suffices to have {ap, ip} together with either tr or nf to have all four as in $\le_{bo}$.

6.6.3 The Penguin Problem

It has become traditional to see how theoretical results can be applied to the following problem: "birds fly", "penguins (P) are birds (B)", and "penguins don't fly (F)". What are the implications of these three statements with respect to the various deductive relations?

It is important to distinguish universally true statements such as "penguins are birds" from empirical statements that merely have a probablility of being true such as "birds fly". "Penguins are birds" is a category statement, a definition of "penguin", which can be expressed in terms of Boolean deduction: $p \le B$. [This has many equivalent ways to be expressed including pB=p, pB'=0, p v B = B, B' $\le$ p', p' v B = 1, and (B|p) = (1|p), any of which can be used to simplify the remaining conditionals.] Alternately, in the finite case, this can be equivalently expressed as P(B|p) = 1, the probability of a bird given a penguin is 1. The 3rd statement "penguns don't fly" is the non-observation of a penguin that flies. It is an empirical statement.

If "birds fly" is interpreted as "all birds fly", then by Boolean logic, $p \le B$ and $B \le F$ would imply that $p \le F$, that all penguins fly, which would make the conditional " (F'|P) that penguins don't fly equal to (0|p), a contradiction given a penguin. So "Birds fly" will be considered a partially true conditional (F|B) - given a bird, it flies.

Representing these three statements as (F|B), (B|p) and (F'|p), respectively, then J = {(F|B), (B|p), (F'|p)}. But (B|p) and (F'|p) have the same condition, and so their conjunction, (BF'|p), implies each of them with respect to all deductive relations in the hierarchy including $\le_{bo}$. So the DCS with respect to $\le_x$ generated by J equals H_x{(F|B), (BF'|p)}.

The set of 3 combinations of these two is the conjunctive closure C(J):

$$C(J) = \{(F|B), (BFp'|B \vee p), (BF'|p)\}.$$

Here, $(BFp'|B \lor p)$ represents the event of a flying, non-penguin bird given it is a bird or a penguin.

However, by definition every penguin is a bird, that is, $p \leq B$. Therefore $(B \lor p) = B$, and so $(BFp'|B \lor p)$ becomes $(BFp'|B)$, which implies $(F|B)$ with respect to all deductive relations in the hierarchy including $\leq_{bo}$. Therefore

$$C(J) = \{(BFp'|B), (BF'|p)\}.$$

Thus we are left with "given a bird, it is a flying non-penguin" and "given a penguin, it is a non-flying bird."

By Theorem 6.4.13, for $x = $ ip, nf or $\land$, the DCS generated by J with respect to $\leq_x$ is just the principal DCS generated by the conjunction of the conditionals of J. Thus $H_x(J) = H_x\{(BFp'|B)(BF'|p)\} = H_x(BFp'|B \lor p) = H_x(BFp'|B)$ since $(B \lor p) = B$.

Thus, using the results of Sections 6.4.6 – 6.4.8,

$$H_{ip}(BFp'|B) = \{(x \mid By): \text{any } x, y \in \mathcal{B}\}.$$

and,

$$H_{nf}(BFp'|B) = \{(BFp' \lor B' \lor x \mid y): \text{any } x, y \in \mathcal{B}\}.$$

and,

$$H_\land(BFp' \mid B) = \{(BFp' \lor x \mid By): \text{any } x, y \in \mathcal{B}\},$$

The latter being the set of all conditionals whose condition is a subset of B (birds) and whose conclusion includes BFp', the non-penguin flying birds.

Concerning the other deductive relations, except for $x = \lor$ or tr, by Theorem 6.4.10.

$$H_x(J) = H_x(BFp' \mid B) \cup H_x(BF'|p)$$

[Note that even though the conditional, $(BFp' \mid B \lor p)$, is equal to the conjunction, $(F|B) \land (BF'|p)$, of the other two generators, in general it is not in the principal DCS's of either of them. It is only because (B ∨ p) = B that (BFp' | B ∨ p) can be dropped from C(J).]

$H_x(J)$ is the union of the principal DCS generated by (BFp' | B), the "flying, non-penguins given a bird", and the principal DCS generated by (BF'|p), the "non-flying birds given a penguin".

Let x = ap.
So $H_{ap}(J) = H_{ap}(BFp'|B) \cup H_{ap}(BF'|p)$. This is just all conditionals whose condition is either a superset of B or of p. As previously mentioned, these generate the conditional zeroes (0|B) and (0|p).

Let x = pm. Then applying Section 6.4.6 – 6.4.8 and noting that p ≤ B also means that $pB' = 0$, $pB = p$, and $B' \le p'$, then

$$H_{pm}(J) = \{(BFp' \lor B' \lor x \mid BFp' \lor y): \text{any } x, y \in \mathcal{B}\}$$
$$\cup \, \{(F'p \lor p' \lor x \mid F'p \lor y): \text{any } x, y \in \mathcal{B}\}.$$

Using streamlined bracket notation [BFp' ∨ B'] for the Boolean ideal (DCS) generated by the ordinary Boolean proposition (BFp' ∨ B'), $H_{pm}(J)$ can be written more concisely as

$$H_{pm}(J) = ([BFp' \lor B'] \mid [BFp']) \cup ([F'p \lor p'] \mid [F'p])$$

Signifying all conditionals whose premise includes BFp' and whose conclusion includes (BFp' ∨ B') together with all conditionals whose premise includes F'p and whose conclusion includes (F'p ∨ p'). Note that $F'p \lor p' = F' \lor p' = (Fp)'$, the complement of a flying Penguin.

Thus $H_{pm}(J)$ consists of the union of two sets of conditionals: The first set consists of all conditionals whose condition includes the flying non-penguin birds, and whose conclusion includes the flying non-penguins and the non-birds. The second set of conditionals consists of all those whose context includes the non-flying penguins and whose conclusion includes (Fp)', the complement of the flying penguins.

Let x = m∧. If p ≤ B, then

$$H_{m\wedge}(J) = \{(BFp' \vee x \mid BFp' \vee By): \text{any } x, y \in \mathcal{B}\}$$
$$\cup \{(F'p \vee x \mid F'p \vee yp): \text{any } x, y \in \mathcal{B}\}.$$

So the implications of J with respect to $\leq_{m\wedge}$ are all conditionals whose condition is between BFp' and B, and whose conclusion includes BFp' together with all conditionals whose condition is between $F'p$ and p, and whose conclusion includes $F'p$.

Let x = mv. Then $H_{mv}(J) = H_{mv}(BFp' \mid B) \cup H_{mv}(BF'p \mid p)$
$$= \{(BFp' \vee B' \vee x \mid B \vee y): \text{any } x, y \in \mathcal{B}\}$$
$$\cup \{(F'p \vee p' \vee x \mid p \vee y): \text{any } x, y \in \mathcal{B}\}.$$

Thus $H_{mv}(J)$ is the set of all conditionals whose condition includes the birds and whose conclusion includes the non-birds and the flying non-penguin birds, together with all conditionals whose condition includes the penguins and whose conclusion includes $F'p \vee p'$, the complement of the flying penguins.

Let x = bo. Again by the results of Section 6.4.8,

$$H_{bo}(J) = H_{bo}(BFp' \mid B) \cup H_{bo}(BF' \mid p).$$

$$H_{bo}(J) = \{(BFp' \vee x \mid B): \text{any } x \in \mathcal{B}\} \cup \{(BF'p \vee x \mid p): \text{any } x \in \mathcal{B}\}.$$

In streamlined notation, this is $H_{bo}(J) = ([Fp'] \mid B) \cup ([F'] \mid p)$

That is, the Boolean implications of J are all conditionals whose condition is B (bird) and whose conclusion includes the flying non-penguin birds and together with all conditionals whose condition is p (penguin) and whose conclusion includes the non-flyers, using that $pB = p$.

In other words, the implications of J with respect $\leq_{bo}$ are the conditionals "a flying non-penguin if a bird" together with "a non-flying bird if a penguin".

For completeness, let x = tr. In general,

$$H_{tr}(J) = H_{tr}\{(BFp'|B), (BF'|p)\}.$$

The Deductive Extension Theorem 6.4.10 does not apply for $\leq_{tr}$. But by Theorem 6.5.1,

$$H_{tr}(J) = H_{tr}\{(BFp')(B)(BF')(p)\} = H_{tr}\{0\} = \mathcal{B}/\mathcal{B},$$

which is the whole conditional event algebra. Thus, with respect to $\leq_{tr}$, J generates a universal contradiction. "Birds fly" is not completely true.

Let x = v. So
$$H_v(J) = H_v\{(BFp'|B), (BF'|p)\}.$$

According to Theorem 6.5.6, $H_v\{(a|b), (c|d)\} = H_v(a|b) \cup H_v(c|d) \cup H_v(abcd \mid b \vee d)$. Therefore, $H_v(J) = H_v(BFp'|B) \cup H_v(BF'|p) \cup H_v((BFp')(B)(BF')p \mid B \vee p)$. But $H_v((BFp')(B)(BF')p \mid B \vee p) = H_v(0 \mid B \vee p) = H_v(0 \mid B)$.

Since $(0 \mid B) \leq_v (BFp'|B)$ therefore $H_v(BFp'|B) \subseteq H_v(0 \mid B)$. Furthermore, $(0 \mid B) \leq_v (BF'|p)$ because $(0)(B) \leq (BF')(p)$ and $(0 \vee B' \leq (BF') \vee p'$, the latter inequality because $B' \leq p'$. Therefore $H_v(BF'|p) \subseteq H_v(0|B)$.

Thus $H_v(J) = H_v(0|B) = \{(x \mid B \vee y): \text{any } x, y \text{ in } \mathcal{B}\}$. That is, with respect to deductive relation $\leq_v$, J generates all conditionals whose premise includes B.

Only Extended Boolean Deduction $\leq_{bo}$ seems to fully capture the essence of the initial statements and the implications of a set of uncertain conditionals. However, the solution to a practical problem with partially true conditionals usually requires statements about the conditional probabilities of those conditioinals. In a later section this aspect will be added using the maximum entropy approach to estimating conditional probabilities.

6.6.4 Absent-Minded Coffee Drinker Example

The second example by H. Pospesel [37] is a typical "expert system" inference problem called the "absent-minded coffee drinker": "Since my spoon is dry I must not have sugared my coffee, because the spoon would be wet if I had stirred the coffee, and I wouldn't have stirred it unless I had put sugar in it."

Let D denote "my spoon is dry"; let G denote "I sugared my coffee"; and let R denote "I stirred my coffee". Translating into this terminology the set of premises is J = {D, (D'|R), (R'|G')}. Therefore the conjunctive closure $C(J)$ = {D, (D'|R), (R'|G'), (D)(D'|R), (D)(R'|G'), (D'|R)(R'|G'), (D)(D'|R)(R'|G')}.

In the Boolean 2-valued logic, the implications of J are those of the conjunction D(D'|R)(R|G) where the conditionals are equated to their material conditionals and have a conjunction D(D' ∨ R')(R' ∨ G) = DR', that the spoon is dry and the coffee is not stirred.

More generally using the operations on uncertain conditionals the conjunctive closure of J is $C(J)$ = {D, (D'|R), (R'|G'), DR', DG ∨ DR'G', (D'RG ∨ R'G' | R ∨ G'), DR'}. Obviously, the propositions D and DR' (twice) are implications with respect to all deductive relations of DR', and so for all deductive relations $\leq_x$ their implications are included in H_x{(D'|R), (R'|G'), DR', DG ∨ DR'G', (D'RG ∨ R'G' | R ∨ G')}. Furthermore, (DG ∨ DR'G') = D(G ∨ R'G') = D(G ∨ R') = DG ∨ DR'), which is thus also an implication of DR'. So dropping (DG ∨ DR'G') from J, $H_x(J)$ = H_x{(D'|R), (R'|G'), (D'RG ∨ R'G' | R ∨ G'), DR'}.

Now, according to the Corollary 6.4.12, for any of the elementary deductive relations $\leq_X$ or their combinations, except for $\leq_{tr}$ and $\leq_v$, $H_x(J)$ = H_x(D'|R) ∪ H_x(R'|G') ∪ H_x(D'RG ∨ R'G' | R ∨ G') ∪ H_x(DR').

For x = ip, nf or ∧, by Theorem 6.4.13, $H_x(J)$ = H_x(D(D'|R)(R'|G')) = H_x(DR'). Therefore $H_{nf}(J)$ = H_{nf}(DR') = {(DR' ∨ Y | Z): any Y, Z in B}. That is, the implications of J when its conditionals are regarded as non-false, are all those conditionals with any condition and whose conclusion includes the event DR', that "my spoon is dry" and "I did not stir my coffee". Notice that G', "I did not sugar my coffee", is not an implication of J with respect to the non-falsity deductive relation, and neither is it a valid consequence of J in the 2-valued Boolean logic. In the 2-valued logic the implications of J are the universally conditioned events that include DR', that the spoon is dry and my coffee is not stirred. But the implications with respect to the "non-falsity" deductive relation $\leq_{nf}$ include all those with any other condition attached.

Similarly, by Theorem 6.4.13, $H_\wedge(J) = H_\wedge(DR') = \{(DR' \vee Y \mid Z):$ any Y, Z in $B\} = H_{nf}(J)$, and so in this case the implications with respect to $\leq_\wedge$ are equal to the implications with respect to $\leq_{nf}$.

Turning to $\leq_{pm}$, there are no further simplifications to the union of the four sets. So $H_{pm}(J) = H_{pm}(D'|R) \cup H_{pm}(R'|G') \cup H_{pm}(DR') \cup H_{pm}(D'RG \vee R'G' \mid R \vee G')$. By the results of Section 6.4.6, each of these can be expressed in terms of Boolean deductions with respect to the conditions and conclusions, the "if" and the "then" parts of these four conditionals.

The set of implications with respect to $\leq_{pm}$ of $J = \{D, (D' \mid R), (R'|G')\}$ is the union of four sets of conditionals. By Section 6.4.6, $H_{pm}(D'|R)$ is the set of conditionals whose condition includes my spoon being non-dry and my coffee stirred and whose conclusion includes $(DR)'$, the negation of a dry spoon and stirred coffee. $H_{pm}(R'|G')$ is the set of all conditionals whose condition includes my not stirring nor sugaring my coffee and whose conclusion includes sugaring my coffee or not sugaring nor stirring it. $H_{pm}(DR')$ is the set of all conditionals whose condition and conclusion include my not stirring my coffee and my spoon being dry. $H_{pm}(D'RG \vee R'G' \mid R \vee G')$ is all those conditionals whose condition includes my stirring and sugaring my coffee or not sugaring my coffee and whose conclusion includes my stirring and sugaring my coffee and wetting my spoon or neither stirring nor sugaring my coffee.

All the conditionals in $H_{pm}(D'|R)$ have conditional probability no less than $P(D'|R)$. Those in $H_{pm}(R'|G')$ have conditional probability no less than $P(R'|G')$. Those in $H_{pm}(DR')$ have conditional probability no less than $P(DR')$, and those conditionals in $H_{pm}(D'RG \vee R'G' \mid R \vee G')$ have conditional probability no less than $P(D'RG \vee R'G' \mid R \vee G')$.

This problem well illustrates the need to utilize additional constraints on the conditionals to narrow the range of implications (solutions) to the collection of conditional statements. This can be done by imposing bounds on the conditional probabilities of these conditional statements. But even this is not enough. This problem will be addressed in terms of the concepts of conditional independence and maximum entropy in a subsequent section describing the work of Wilhelm Rödder [157, 125] who uses this algebra and conditional independence to come to a rapid solution based upon maximum information entropy.

To simplify the situation with our absent-minded coffee drinker, if the spoon is observed to be dry (D=1) then $H_{pm}(J) = H_{pm}(0|R) \cup H_{pm}(R'|G') \cup H_{pm}(R') \cup H_{pm}(R'G' | R \vee G')$. Since $(0|R) \leq_{pm} (R'|G')$ is true, therefore $H_{pm}(0|R) \supseteq H_{pm}(R'|G')$, and since $(0|R) \leq_{pm} R'$ is true, therefore $H_{pm}(0|R) \supseteq H_{pm}(R')$, and since $(0|R) \leq_{pm} (R'G' | R \vee G')$ is true, therefore $H_{pm}(0|R) \supseteq H_{pm}(R'G'|R \vee G')$. Thus $H_{pm}(J)$ reduces to a single set of conditionals. $H_{pm}(J) = H_{pm}(0|R) = \{(0 \vee R' \vee Y | 0 \vee Z)$: any $Y, Z\} = \{(R' \vee Y | Z)$: any events $Y, Z\}$.

So if I observe my spoon to be dry, the implications of J with respect to $\leq_{pm}$ are just all conditionals whose conclusion includes R', that I didn't stir my coffee. There are no other implications. Thus the probability that I sugared my coffee is not further constrained.

Finally consider the implications with respect to the deductive relation $\leq_{bo}$. From $H_x(J) = H_x(D'|R) \cup H_x(R'|G') \cup H_x(DR') \cup H_x(D'RG \vee R'G' | R \vee G')$ it follows that $H_{bo}(J) = H_{bo}(D'|R) \cup H_{bo}(R'|G') \cup H_{bo}(DR') \cup H_{bo}(D'RG \vee R'G' | R \vee G') = \{(D'R \vee Y | R)$: and Y in B$\} \cup \{(R'G' \vee Y | G')$: any Y in B$\} \cup \{(DR' \vee Y)$: any Y in B$\} \cup \{(D'RG \vee R'G' \vee Y | R \vee G')$: any Y in B$\}$. Again, there is too much freedom on the conditionals for a practice solution.

6.6.5 Absent-Minded Coffee Drinker Revisited

It interesting to see what happens with this example when the conditional (R'|G') in J is replaced by (R|G), which is the contra-positive of its logical converse. Instead of saying "I wouldn't have stirred my coffee unless I had put sugar in it" suppose it was "if I sugared my coffee then I stirred it." Thus $J = \{D, (D'| R), (R|G)\}$.

In the Boolean 2-valued logic, the implications of J are those of the conjunction D(D'|R)(R|G) where the conditionals are equated to their material conditionals and have a conjunction $D(D' \vee R')(R \vee G') = DR'G'$. In this case, the implications would include that the coffee is the not sugared or stirred and the spoon is dry.

More generally the conjunctive closure of J is $C(J) = \{D, (D'|R), (R|G), D(D'|R), D(R|G), (D'|R)(R|G), D(D'|R)(R|G)\} = \{D, (D'|R), (R|G), DR', DG' \vee DRG, (D'RG' \vee D'RG | R \vee G), DR'G'\}$. Obviously, the propositions D and DR' are implications of DR'G' with respect to all deductive relations, and so for all deductive relations $\leq_x$

their implications are included in $H_x(J) = H_x\{(D'|R), (R|G), (DG' \vee DRG), (D'R | R \vee G), (DR'G')\}$. Furthermore, $(DG' \vee DRG) = D(G' \vee RG) = D(G' \vee R) = D(R'G' \vee R) = DR'G' \vee DR$, which thus is also an implication of $DR'G'$. So dropping $(DG' \vee DRG)$ from J, $H_x(J) = H_x\{(D'|R), (R|G), (D'R | R \vee G), (DR'G')\}$.

Note that the proposition $DR'G'$ (having a dry spoon, unstirred coffee, and un-sugared coffee) which is the conjunction of the three original conditionals of $J = \{D, (D' | R), (R|G)\}$, is an implication with respect to all these deductive relations. It is a logical consequence of J. Furthermore, by rearranging the conditioning, its probability $P(DR'G') = P(D)P(R'G'|D) = P(D)P((G'|R')|D)P(R'|D) = P(D)P(G'|DR')P(R'|D)$. This latter product has easily estimated conditionals probabilities: $P(D) = 1$ by observation, and both $P(G'|DR')$ and $P(R'|D)$ are also close to or equal to 1. This is one way the reasoning can go since the initial phrasing was in terms of conditionals whose probabilities are easily estimated.

Furthermore, $(D'R | R \vee G) \leq_{pm} (D'|R)$ because $(D'R)(R \vee G) \leq (D'R)$ and $D'R \vee (R \vee G)' \leq D'R \vee R'$. Thus $H_{pm}(J) = H_{pm}(R|G) \cup H_{pm}(D'R | R \vee G) \cup H_{pm}(DR'G')$. So $H_{pm}(J) = \{(RG \vee G' \vee Y | RG \vee Z)$: any Y, Z in $B\} \cup \{(D'R \vee R'G' \vee Y | D'R \vee Z)$: any Y, Z in $B\} \cup \{(DR'G' \vee Y | DR'G' \vee Z)$: any Y, Z in $B\}$. The conditionals in $H_{pm}(R|G)$ all have conditional probability no less than $P(R|G)$, and similarly for the conditionals in $H_{pm}(D'R | R \vee G)$ and in $H_{pm}(DR'G')$.

If the spoon is observed to be dry (D=1), then the 3 conditionals $\{(R|G), (D'R | R \vee G), (DR'G')\}$ reduce to $\{(R|G), (0 | R \vee G), DR'G'\}$. In that case, $(0 | R \vee G) \leq_{pm} (R|G)$ since $(0)(R \vee G)' \leq RG$ and $0 \vee (R \vee G)' \leq R \vee G' = R \vee R'G'$. So $(R|G)$ can be dropped from the generating set J leaving $\{(0 | R \vee G), DR'G'\}$.

By similar methods, the implications of J with respect to $\leq_{nf}$ are $H_{nf}(J) = \{(DR'G' \vee Y | Z)$: any Y, Z in $B\}$, namely any conditionals whose conclusion includes $DR'G'$.

Finally, with respect to $\leq_{bo}$, the four conditional propositions in $J = \{(D'|R), (R|G), (D'R | R \vee G), DR'G'\}$ have non-equivalent conditions and so cannot imply one another with respect to $\leq_{bo}$. It follows that $H_{bo}(J) = H_{bo}(D'|R) \cup H_{bo}(R|G) \cup H_{bo}(D'R | R \vee G) \cup H_{bo}(DR'G') = \{(D'R \vee Y | R)$: any Y in $B\} \cup \{(RG \vee Y | G)$: any Y in $B\} \cup \{(D'R \vee Y | R \vee G)$: any Y in $B\} \cup \{(DR'G' \vee Y)$: any Y in $B\}$. So the implications with respect to $\leq_{bo}$ include $H_{bo}(D'|R)$, all those conditionals with the condition that I stirred my coffee (R) and with a conclusion that includes a non-dry spoon and stirred coffee (D'R). $H_{bo}(R|G)$ is all conditionals with sugared coffee (G) as condition and with a conclusion that includes RG, stirred and sugared coffee. $H_{bo}(D'R | R \vee G)$ is all conditionals with conclusions that include D'R and with condition $R \vee G$, of either stirred coffee or sugared coffee. $H_{bo}(DR'G')$ is simply the set of all (universally unconditioned) events that include DR'G', a dry spoon and unstirred, un-sugared coffee. All of these conditionals have probabilities no less than the corresponding conditional that generates them.

6.7 Simplest Finite Deductively Closed Sets of Conditionals

The simplest examples of deductively closed sets of conditionals with respect to various deductive relations are based on the simplest Boolean algebras B having the smallest number of atoms. The simplest examples of deductively closed sets of conditionals with respect to various deductive relations will be based on the simplest Boolean algebras B having the smallest number of atoms.

Generating set B with zero atoms: Only the zero element, 0, has no atoms. One proposition, {0} and one conditional ({0} | {0}), abbreviated 0 and (0|0) respectively, the latter also denoted U. (Strictly speaking a Boolean algebra B must have at least one atom and two members.)

6.7.1 Generating with a Boolean Algebra B with One Atom

The 2-element Boolean algebra {0,1}. Two propositions usually abbreviated as 0 and 1. Three conditionals: B/B = {(0|1), (1|1), and (1|0)}, where (0|0) = (1|0). These are usually abbreviated 0, 1, U, with U interpreted as "inapplicable" or "undefined".

a) Deductive Relation $\leq_{ap}$.

Since the conditional propositions 0 and 1 are applicability equivalent, that is, since (0 $=_{ap}$ 1), they generate the same deductively closed set, namely $H_{ap}(0) = H_{ap}(1) = \{(x|y): (0|1) \leq_{ap} (x|y)\} = \{(x|y): 1 \leq y)\} = \{0, 1\}$. The conditional U generates its own DCS, $H_{ap}(U)$ since U is applicability equivalent only to itself.

$H_{ap}(U) = \{(x|y): (1|0) \leq_{ap} (x|y)\} = \{(x|y): 0 \leq y\} = \{0, 1, U\} = \mathcal{B}/\mathcal{B}$.

b) Deductive Relation $\leq_{ip}$

Since $(0 =_{ip} 1)$ also holds for $\leq_{ip}$, and U is again inapplicability equivalent only to itself, preorder $\leq_{ip}$ has the same generators as does preorder $\leq_{ap}$. Two conditional propositions are equivalent in applicability or inapplicability if and only if they have equivalent conditions. But $\leq_{ap}$ and $\leq_{ip}$ generate different DCS's:

$H_{ip}(1) = \{(x|y): (1|1) \leq_{ip} (x|y)\} = \{(x|y): y \leq 1\} = \{0, 1, U\} = \mathcal{B}/\mathcal{B}$
$H_{ip}(U) = \{(x|y): (1|0) \leq_{ip} (x|y)\} = \{(x|y): y \leq 0\} = \{U\}$

c) Deductive Relation $\leq_{nf}$

Since $(1 =_{nf} U)$, U and 1 generate the same DCS, namely $H_{nf}(1) = H_{nf}(U) = \{1, U\}$. However,

$H_{nf}(0) = \{(x|y): (0|1) \leq_{nf} (x|y)\} = \{(x|y): 0 \leq x \vee y'\} = \mathcal{B}/\mathcal{B} = \{0,1,U\}$.

d) Deductive Relation $\leq_{tr}$

Since $(0 =_{tr} U)$, 0 and U generate the same DCS, namely $H_{tr}(0) = H_{tr}(U) = \{(x|y): (0|1) \leq_{tr} (x|y)\} = \{(x|y): 0 \leq xy\} = \mathcal{B}/\mathcal{B} = \{0,1,U\}$.
$H_{tr}(1) = \{(x|y): (1|1) \leq_{tr} (x|y)\} = \{(x|y): 1 \leq xy\} = \{1\}$.

So the atom 1 generates its own singleton DCS.

Thus in summary concerning the elementary preorders, $H_{ap}(0) = H_{ap}(1) = \{0, 1\}$ and with respect to $\leq_{ap}$ both 0 and 1 are generators. U generates $\mathcal{B}/\mathcal{B}$. $H_{ip}(U) = \{U\}$ while with respect to $\leq_{ip}$ both 0 and 1 generate $\mathcal{B}/\mathcal{B}$. $H_{nf}(1) = H_{nf}(U) = \{1, U\}$ while 0 generates $\mathcal{B}/\mathcal{B}$. $H_{tr}(1) = \{1\}$ while both 0 and U generate $\mathcal{B}/\mathcal{B}$.

6.7.2 Combinations of Elementary Preorders

The conjunction theorem for deductive relations allows us to determine most of the DCS's for combination preorders in the hierarchy by simply taking the intersection of DCS's generated by the individual elementary preorders or generated by the constituent deductive relations. The others can be determined using the Deductive Extension Theorem or Theorems 6.5.5 and 6.5.6.

The DCS's of Conditionals for the 3-element Conditional Event Algebra $\mathcal{B}/\mathcal{B} = \{(0|1), (1|1), (1|0)\} = \{0,1,U\}$ are as follows:

a) Deductive Relation $\leq_{ec}$

By Theorem 6.4.7 on the Prinicipal DCS's with respect to the non-elementary deductive relations,

$H_{ec}(1) = H_{ap \cap ip}(1) = H_{ap}(1) \cap H_{ip}(1) = \{0,1\} \cap \mathcal{B}/\mathcal{B} = \{0,1\}$.
$H_{ap \cap ip}(0) = H_{ap}(0) \cap H_{ip}(0) = H_{ap}(1) \cap H_{ip}(1) = \{0,1\}$.
$H_{ap \cap ip}(U) = H_{ap}(U) \cap H_{ip}(U) = \mathcal{B}/\mathcal{B} \cap \{U\} = \{U\}$.

So $\leq_{ec}$ generates three deductively closed sets including the whole space $\mathcal{B}/\mathcal{B}$. Furthermore, with respect to $\leq_{ec}$, $\mathcal{B}/\mathcal{B}$ is already a finite, non-principal DCS since it is not generated by a single conditional but instead requires both 0 and U. This is another difference from the finite Boolean situation, where all finite DCS's are principal.

b) Deductive Relation $\leq_v$
Using the Theorem 6.4.7 for deductive relations, $\leq_v = \leq_{ap \cap tr}$.

$H_v(1) = H_{ap}(1) \cap H_{tr}(1) = \{0,1\} \cap \{1\} = \{1\}$.
$H_v(0) = H_{ap}(0) \cap H_{tr}(0) = \{0,1\} \cap \{0,1,U\} = \{0,1\}$.
$H_v(U) = H_{ap}(U) \cap H_{tr}(U) = \{0,1,U\} \cap \{0,1,U\} = \mathcal{B}/\mathcal{B}$.

c) Deductive Relation $\leq_\wedge$ Similarly,

$H_\wedge(1) = H_{nf \cap ip}(1) = H_{nf}(1) \cap H_{ip}(1) = \{1,U\} \cap \{0,1,U\} = \{1,U\}$.
$H_\wedge(U) = H_{nf \cap ip}(U) = H_{nf}(U) \cap H_{ip}(U) = \{1,U\} \cap \{U\} = \{U\}$.
$H_{nf \cap ip}(0) = H_{nf}(0) \cap H_{ip}(0) = \{0,1,U\} \cap \{0,1,U\} = \{0,1,U\} = \mathcal{B}/\mathcal{B}$.

d) Deductive Relation $\leq_{pm}$ Similarly,

$H_{pm}(1) = H_{tr \cap nf}(1) = H_{tr}(1) \cap H_{nf}(1) = \{1\} \cap \{1,U\} = \{1\}$.
$H_{pm}(0) = H_{tr \cap nf}(0) = H_{tr}(0) \cap H_{nf}(0) = \{0,1,U\} \cap \{0,1,U\} = \mathcal{B}/\mathcal{B}$.

$H_{pm}(U) = H_{tr \cap nf}(U) = H_{tr}(U) \cap H_{nf}(U) = \{0,1,U\} \cap \{1,U\} = \{1,U\}$.

e) Deductive Relation $\leq_{mv}$ Since $\leq_{mv} = \leq_{pm \cap v}$,

$H_{mv}(1) = H_{pm}(1) \cap H_v(1) = \{1\} \cap \{1\} = \{1\}$.
$H_{mv}(0) = H_{pm}(0) \cap H_v(0) = \{0,1,U\} \cap \{0,1\} = \{0,1\}$.
$H_{mv}(U) = H_{pm}(U) \cap H_v(U) = \{1,U\} \cap \{0,1,U\} = \{1,U\}$.

Since B/B is not generated by any single one of its elements, B/B is a non-principal DCS with respect to $\leq_{mv}$.

f) Deductive Relation $\leq_{m\wedge}$ Since $\leq_{m\wedge} = \leq_{pm \cap \wedge}$,

$H_{m\wedge}(1) = H_{pm}(1) \cap H_\wedge(1) = \{1\} \cap \{1,U\} = \{1\}$.
$H_{m\wedge}(0) = H_{pm}(0) \cap H_\wedge(0) = \{0,1,U\} \cap \{0,1,U\} = \{0,1,U\}$.
$H_{m\wedge}(U) = H_{pm}(U) \cap H_\wedge(U) = \{1,U\} \cap \{U\} = \{U\}$.

In addition $H_{m\wedge}\{1,U\} = H_{pm}\{1,U\} \cap H_\wedge\{1,U\} = \{1,U\} \cap \{1,U\} = \{1,U\}$. That is, $\{1,0\}$ is a non-principal DCS with respect to $\leq_{m\wedge}$ since it requires two propositions to generate it.

g) Deductive Relation $\leq_{bo}$ The Boolean deductive relation $\leq_{bo} = \leq_{ec \cap tr}$. So as expected

$H_{bo}(1) = H_{ec \cap tr}(1) = H_{ec}(1) \cap H_{tr}(1) = \{0,1\} \cap \{1\} = \{1\}$.
$H_{bo}(0) = H_{ec \cap tr}(0) = H_{ec}(0) \cap H_{tr}(0) = \{0,1\} \cap \{0,1,U\} = \{0,1\}$
$H_{bo}(U) = \{U\}$ since only U has 0 condition. Finally,
$H_{bo}\{1, U\} = H_{ec}\{1,U\} \cap H_{tr}\{1, U\} = \{0,1,U\} \cap \{0,1,U\} = B/B$.

i) Deductive Relation $\leq_1$ Only equal conditionals satisfy $\leq_1$ and so the DCS's with respect to $\leq_1$ are just the equivalence classes of equal conditionals, namely $\{0\}$, $\{1\}$ and $\{U\}$ each generated by its single conditional. The subsets $\{0,1\}$, $\{0,U\}$ and $\{1, U\}$ are also DCS's since the conjunction of any two of the three conditionals 0, 1, or U gives back one of the two components. So these subsets are closed under conjunction and they have only themselves as deductions with respect to $\leq_1$. The whole space $\{0,1,U\}$ is also a DCS requiring all three of its conditionals to generate it.

j) Deductive Relations $\leq_0$ All conditionals are deducible from any one conditional with respect to $\leq_0$. So there is just one DCS with respect to $\leq_0$, namely the whole space $\{0,1,U\}$ and it is generated by any of its members. This is true no matter what the original Boolean algebra $\mathcal{B}$.

This completes the deductively closed sets of conditionals of $\mathcal{B}/\mathcal{B} = \{0,1,U\}$ with respect to the 13 deductive relations identified in the hierarchy.

It is not so remarkable that all 13 of these preorders yield different collections of DCS's for this simplest case since even the 3-element Conditional Logic $\mathcal{B}/\mathcal{B}$ has 8 subsets. Each preorder determines which of these 8 subsets will form a deductively closed set with respect to it. So there are potentially $2^8 = 256$ different possible choices of a subset of the 8 to be a DCS. These are subsets of subsets of the 3 original conditionals - the so-called second order predicates.

6.7.3 The DCS's of the 9-Element Conditional Event Algebra $\mathcal{B}/\mathcal{B}$

Consder the deductively closed sets of the 9-Element conditional event algebra $\mathcal{B}/\mathcal{B}$ of the 4-element Boolean Algebra $\mathcal{B} = \{0, b, b', 1\}$ generated by a single non-trivial event b.

$$\mathcal{B}/\mathcal{B} = \{(0|1), (b|1), (b'|1), (1|1), (0|b), (1|b), (0|b'), (1|b'), (0|0)\}$$
$$= \{0, b, b', 1, (0|b), (1|b), (0|b'), (1|b'), U\}.$$

There are $2^9 = 512$ different subsets of $\mathcal{B}/\mathcal{B}$, and a particular deductive relation will make some collection of these subsets a DCS with respect to it. There are therefore 2^{512} different collections of subsets of conditionals each of which is a possible member of the collection of all DCS's generated by a given deductive relation, and that is just for the 4-Element Boolean algebra. Deduction from conditional information is indeed complex!

The Four Elementary Deductive Relations $\leq_{ap}, \leq_{tr}, \leq_{nf},$ and $\leq_{ip}$

Nine Conditionals of $\mathcal{B} = \{0, b, b', 1\}$

Preorders	DCS #	1	b	b'	(1|b)	(1|b')	0	(0|b)	(0|b')	U
$\leq_{ap}$	1	G	G	G			G			
	2	+	+	+	G		+	G		
	3	+	+	+		G	+		G	
	4	+	+	+	+	+	+	+	+	G
	5	+	+	+	G	H	+	G	H	
$\leq_{tr}$	1	+	+	+	+	+	G	G	G	G
	2	G								
	3	+	G		G					
	4	+		G		G				
$\leq_{nf}$	1	+	+	+	+	+	G	+	+	+
	2	G			G	G				G
	3	+	G		+	+			G	+
	4	+		G	+	+		G		+
$\leq_{ip}$	1	G	G	G	+	+	G	+	+	+
	2					G		G		+
	3						G		G	+
	4									G

Table 6.7.1 Deductively Closed Sets of $\mathcal{B}/\mathcal{B} = \{0, b, b', 1, (0|b), (1|b), (0|b'), (1|b'), U\}$ with respect to the preorders (deductive relations) $\leq_{ap}, \leq_{tr}, \leq_{nf},$ and $\leq_{ip}$

Key: # - Numbered deductively closed set (DCS) for a preorder
 + - Included in the DCS of that row
 G, H - Generators of the DCS of that row; one of each present in a row is required to generate that row

The DCS's in the above table are determined by methods similar to those used to determine the DCS's of the 2-element Boolean algebra.

For example, for the preorder $\leq_{nf}$ the conditional (b|1), which is b, generates the DCS $H_{nf}(b) = \{(x|y): b \vee 1' \leq x \vee y'\} = \{(x|y): b \leq x \vee y'\}$. If $y = 0$, then $b \leq x \vee y'$ is satisfied. So (0|0) is in $H_{nf}(b)$. If $y = b'$, then the inequality is also satisfied for any x. So (0|b') and (1|b') are also in $H_{nf}(b)$. If $y = b$ then the inequality is satisfied if and only if $b \leq x$. So (1|b) is in $H_{nf}(b)$. Finally, if $y = 1$, then again $b \leq x$, and so (1|1) is in $H_{nf}(b)$. So in Table 4.1 all of the conditionals of DCS #3 under the deductive relation $\leq_{nf}$ have been shown to be in $H_{nf}(b)$, the DCS with respect to $\leq_{nf}$ generated by b. Now $H_{nf}(0|b') = H_{nf}(b)$ since the same inequality shows up. That is, $H_{nf}(0|b') = \{(x|y): 0 \vee (b')' \leq x \vee y'\} = \{(x|y): b \leq x \vee y'\}$. So (0|b') generates the same DCS. Similar examination of the other conditionals generated by b with respect to $\leq_{nf}$ yields no new conditionals. For instance, (1|b) generates all unity events (1|y) for any y, but they are already included. In fact, (1|b) generates DCS #2, which is a subsystem of DCS #3 of preorder $\leq_{nf}$.

Having determined the DCS's for the elementary preorders, the DCS's for the combination preorders can be determined with the help of the Conjunction Theorem for Deductively Closed Sets with respect to two preorders. For example, using the table, one DCS with respect to $\leq_{pm}$, which is $\leq_{tr \cap nf}$, is determined by intersecting the conditionals in DCS #3 of $\leq_{tr}$ with those in DCS #4 of $\leq_{nf}$. The result is $\{1, (1|b)\}$, the DCS with respect to $\leq_{pm}$ generated by (1|b).

In fact, since the whole space $\mathcal{B}/\mathcal{B}$ is a DCS with respect to any combined preorder, a DCS with respect to one of its component preorders will also be a DCS with respect to the combined preorder by intersection of the DCS with the whole space. This common DCS may in general have different generators with respect to the two preorders, and so they have been included in the tables below.

However, some DCS's, like #11 below of preorder $\leq_v$ are not the intersection of DCS's of more elementary preorders. They were determined with the help of the Theorem on Additional Deductive Information and previously determined principal DCS's.

6.7.4 The DCS's of the Combined Deductive Relations (Preorders)

Using the methods just illustrated the DCS's of the other preorders in the hierarchy were determined and are listed in the following tables:

a) Deductive Relation $\leq_{ec}$

Nine Conditionals of $\mathcal{B} = \{0, b, b', 1\}$

DCS#	1	b	b'	(1|b)	(1|b')	0	(0|b)	(0|b')	U
1	G	G	G			G			
2	G	G	G	H		G	H		
3	G	G	G		H	G		H	
4	G	G	G	H	K	G	H	K	L
5	G	G	G	H	K	G	H	K	
6									G
7				G			G		H
8					G			G	H
9				G			G		
10						G		G	

Table 6.7.2 Deductively Closed Sets of $\mathcal{B}/\mathcal{B} = \{0, b, b', 1, (0|b), (1|b), (0|b'), (1|b'), U\}$ with respect to the deductive relation $\leq_{ec}$

Key: # - Numbered deductively closed sets (DCS's) of the deductive relation $\leq_{ec}$

G, H, K, L - Generators of the DCS of that row; one of each present in a row is required to generate that row

b) Deductive Relation $\leq_V$

Nine Conditionals of $\mathcal{B} = \{0, b, b', 1\}$

DCS#	1	b	b'	(1|b)	(1|b')	0	(0|b)	(0|b')	U
1	+	+	+			G			
2	G								
3	+	G							
4	+								
5	+	+	+	+		+	G		
6	+	+		G					
7	+	+	+		+	+		G	
8	+		+	G					
9	+	+	+	+	+	+	+	+	G
10	+	+	+	+	+	+	J	J	
11	+	+	+	J	J	+			
12	+	+	+		J	J			
13	+	+	+	J		J			

Table 6.7.3 Deductively Closed Sets of $\mathcal{B}/\mathcal{B} = \{0, b, b', 1, (0|b), (1|b), (0|b'), (1|b'), U\}$ with respect to the deductive relation $\leq_V$

Key: # - Numbered deductively closed sets (DCS's) for a preorder
+ - Conditional included in the DCS of that numbered row
G - Generator of the DCS of that row; any G
J - Joint generators; all conditionals with J required

c) Deductive Relation $\leq_{pm}$

Nine Conditionals of $\mathcal{B} = \{0, b, b', 1\}$

DCS#	1	b	b'	(1|b)	(1|b')	0	(0|b)	(0|b')	U
1	+	+	+	+	+	G	+	+	+
2	+			+	+				G
3	+	+		+	+			G	+
4	+		+	+	+		G		+
5	G								
6	+	G		+					
7	+		G						
8	+		G		+				
9	+				G				
10	+			J	J				
11	+	J		+	J				
12	+		J	+	J				
13	+	J		+	+				J
14	+		J	+	+				J

Table 6.7.4 Deductively Closed Sets of $\mathcal{B}/\mathcal{B} = \{0, b, b', 1, (0|b), (1|b), (0|b'), (1|b'), U\}$ with respect to the preorder $\leq_{pm}$

Key: # - Numbered deductively closed sets (DCS's) for a preorder
+ - Conditional included in the DCS of that numbered row
G - Generator of the DCS of that row; any G
J - Joint generators; all conditionals with J required

d) Deductive Relation ≤_∧

Nine Conditionals of $\mathcal{B} = \{0, b, b', 1\}$

DCS#	1	b	b'	(1|b)	(1|b')	0	(0|b)	(0|b')	U
1	+	+	+	+	+	G	+	+	+
2				+		G			+
3					+			G	+
4									G
5	G			+	+				+
6		G							+
7			G						+
8	+	G		+	+			+	+
9	+		G	+	+		+		+

Table 6.7.5 Deductively Closed Sets of $\mathcal{B}/\mathcal{B} = \{0, b, b', 1, (0|b), (1|b), (0|b'), (1|b'), U\}$ with respect to the deductive relation ≤_∧

Key: # - Numbered deductively closed sets (DCS's) for a preorder
      + - Conditional included in the DCS of that numbered row
      G - Generator of the DCS of that row

e) Deductive Relation $\leq_{mv}$

Nine Conditionals of $\mathcal{B} = \{0, b, b', 1\}$

DCS#	1	b	b'	(1|b)	(1|b')	0	(0|b)	(0|b')	U
1	+	+	+			G			
2	G								
3	+	G							
4	+		G						
5	+	+	+	+		J	J		
6	+			G					
7	+	J			J				
8	+		+	+			G		
9	+	+	+		+	J		J	
10	+				G				
11	+	+			+			G	
12	+		J		J				
13	+	+	+	+	+	+	J	J	J
14	+			+	+				G
15	+	+		+	+			J	J
16	+		+	+	+		J		J
17	+	+	+	+	+	+	J	J	
18	+			J	J				
19	+	+		J	+			J	
20	+		+	J	+		J		
21	+	+	+	J	J	J			
22	+	J		J	J				
23	+		J	J	J				
24	+	+	+		J	J			
25	+	+	+	J		J			

Table 6.7.6 Deductively Closed Sets of $\mathcal{B}/\mathcal{B} = \{0, b, b', 1, (0|b), (1|b), (0|b'), (1|b'), U\}$ with respect to the preorder $\leq_{mv}$

Key: # - Numbered deductively closed sets (DCS's) for a preorder
 + - Conditional included in the DCS of that numbered row
 G - Generator of the DCS of that row
 J - Joint generators; all conditionals with J required

f) Deductive Relation $\leq_{m\wedge}$

Nine Conditionals of $\mathcal{B} = \{0, b, b', 1\}$

DCS#	1	b	b'	(1|b)	(1|b')	0	(0|b)	(0|b')	U
1	+	+	+	+	+	G	+	+	+
2				+			G		+
3					+			G	+
4									G
5	G								
6	+	G		+					
7	+		G		+				
8				G					
9					G				
10	J			J					
11	J				J				
12	J			+			J		+
13	J				+		J		+
14	+	J		+	J				
15	+		J	J	+				
16	+	G		G	+			H	+
17	+		G	+	G		H		+
18	+			J	J				
19	+			J	J				J
20	+	J		+	J				J
21	+		J	J	+				J
22				G					G
23					G				G

Table 6.7.7 Deductively Closed Sets of $\mathcal{B}/\mathcal{B} = \{0, b, b', 1, (0|b), (1|b), (0|b'), (1|b'), U\}$ with respect to the deductive relation $\leq_{m\wedge}$

Key: # - Numbered deductively closed sets (DCS's) for a preorder
+ - Conditional included in the DCS of that numbered row
G, H - Generators of the DCS of that row; one of each present in a row is required to generate that row
J - Joint generators; all conditionals with J required

6.7.5 Perspectives on Conditional Logic and Probability

The theory of deductive relations (preorders) for conditionals has been developed and examples have been worked out to illustrate how various deductively closed sets of conditional propositions are generated in a simple initial set of conditionals with respect to the various deductive relations (preorders). Properties of various deductive relations give rise to corresponding properties of the deductively closed sets that they generate. The theory of these deductive systems of conditional propositions has been related to powerdomain theory [158-160, 164].

Hopefully, these very early explorations into a completely unexplored realm of deductive theory may be the basis of further exploration and development by younger mathematicians. The field is wide open for further exploration. Mathematical researchers looking for well-framed, solvable problems will find many here.

6.8 Computations with Conditionals

While the preceding sections provide an adequate theoretical basis for calculating and reasoning with conditional propositions or conditional events, the problem of the complexity of information is no less daunting. Indeed, even without the added computational burden of operating with explicit conditionals, just operating with Boolean expressions in practical situations with, say, a dozen variables, is already too complex for practical pure Bayesian analysis. The reason for this is that in most situations the available information is insufficient to determine a single probability distribution that satisfies the known constraints of the situation. Various possibilities concerning unknown dependences between subsets of variables result in complicated solutions to relatively simple problems.

6.8.1 Pure Bayesian Analysis

For example, consider again the transitivity problem of Section 6.6.1. If "A given B" and "B given C" are both certain, then it follows that "A given C" is also a certainty. But if they are not certain, then by pure Bayesian analysis, $P(A|C)$ can be zero no matter how high are the conditional probabilities of $(A|B)$ and $(B|C)$. This happens because $P(B|C)$ and $P(A|B)$ can be almost 1 while $P(A | BC)$ is zero, and it is the latter probability that appears in the Bayesian solution: $P(A|C) = P(AB \text{ or } AB' | C) = P(AB|C) + P(AB'|C) = P(ABC)/P(C) + P(AB'C)/P(C) = P(ABC | BC) P(BC|C) + P(AB'C | B'C) P(B'C|C) = P(A|BC)P(B|C) + P(A|B'C)P(B'|C)$. Without knowing anything about $P(A|BC)$ or $P(A|B'C)$, nothing more can be said about $P(A|C)$.

6.8.2 A Bayesian Solution Using Maximum Information Entropy

Continuing the example of Section 3.1, knowing that C is true might dramatically change P(A|B) up or down. But if nothing is known one way or the other, the choice of the "maximum information entropy" distribution assumes that P(A|BC) = P(A|B). This latter equation is called the *conditional independence* of A and C given B. It can also be expressed as P(AC|B) = P(A|B)P(C|B) or as P(C|AB) = P(C|B). Using this principle P(A|C) = P(A|B)P(B|C) + P(A|B'C)P(B'|C). So if P(A|B) and P(B|C) are 0.9 and 0.8 respectively, then P(A|C) is at least 0.72. Additionally, since nothing is known one way or the other about the occurrence of A when B is false and C is true, this principle of "maximum indifference" implies that P(A|B'C) should be taken to be ½. So the term P(A|B'C)P(B'|C) contributes (1/2)P(B'|C) = (1/2)(1 − 0.8) = 0.1 to P(A|C) bringing the total to 0.82.

In affect the principle of maximum information entropy chooses that probability distribution P that assumes conditional independence of any two variables that are not explicitly known to have some dependence under the condition. This greatly simplifies computations and often allows situations of several dozen variables to be rapidly analyzed as long as the clusters of dependent variables are not too large and not too numerous. The maximum entropy solution is always one of the possible Bayesian solutions of the situation. If there is just one Bayesian solution, then the two solutions will always agree.

It is a remarkable fact that such a function as the entropy function exists, and it is now clear that it has wide application to information processing under uncertainty. If the n outcomes of some experiment are to be assigned probabilities p_i for i=1 to n subject to some set of constraints, then the distribution of probabilities that assumes conditional independence unless dependence is explicitly known is the one that maximizes the entropy function

$$H(p_1, p_2, p_3 \cdots p_n) = -\sum_{i=1}^{n} p_i \log p_i \qquad (6.8.a)$$

and also satisfies the known constraints. If there is an a priori distribution $q_1, q_2, q_3 \ldots q_n$ then H is given by

$$H(p_1, p_2, p_3 \cdots p_n, q_1, q_2, q_3 \cdots q_n) = -\sum_{i=1}^{n} p_i \log (p_i/q_i) \qquad (6.8.b)$$

This allows maximum entropy updates when additional information is available. See J. E. Shore [165] for a derivation.

W. Rödder [157, 125] and his colleagues at FernUniversität in Hagen, Germany have and are continuing to develop a very impressive interactive computer program SPIRIT that implements this practical approach to the computation of propositions and conditional propositions and their probabilities. Starting with an initially defined set of variables and their values, the user can input statements and conditionals statements about these variables taking various values, and can also assign conditional probabilities to them. The *utility* of having a variable take one of its values can also be incorporated.

6.8.3 Confidence in Maximum Entropy Solutions

While the maximum entropy solution provides the most plausible or "most likely" probability distribution for a situation among all of the Bayesian solutions, it does not immediately provide a means for estimating how much confidence to attach to that solution. This issue has been taken up by E. Jaynes [166], S. Amari [167], and A. Caticha [168].

Jaynes puts the matter as follows: "Granted that the distribution of maximum entropy has a favored status, in exactly what sense, and how strongly, are alternative distributions of lower entropy ruled out?" He proves an entropy "concentration theorem" in the context of an generalized experiment of N independent trials each having n possible results and satisfying a set of m (< n) linearly independent, linear constraints on the observed frequencies of the experiment. Jaynes shows that in the limit as the number N of trials approaches infinity, the fraction F of probability distributions satisfying the m constraints and whose entropy H differs from the maximum by no more than ΔH is given by the Chi-square distribution χ_k^2 with k = n − m −1 degrees of freedom as

$$2N(\Delta H) = \chi_k^2(F) \qquad (6.8.c)$$

That is, the critical, threshold entropy value H_α for which only the fraction α of the probability distributions satisfying the m constraints have smaller entropy is given by

$$H_\alpha = H_{max} - \chi_k^2(1 - \alpha) / 2N \qquad (6.8.d)$$

For N = 1000 independent trials of tossing a 6-sided die and with a significance level $\alpha = 0.05$ and degrees of freedom k = 6 − 1 − 1 = 4, 95% of the eligible probability distributions have entropy no less than

$H_\alpha = H_{max} - 9.49 / 2N = H_{max} - 0.0047$. H_{max} is on the order of 1.7918 for a fair die and 1.6136 for a die with average die value of 4.5 instead of 3.5. Letting $\alpha = 0.005$ it follows that 99.5% of the eligible distributions will have entropy no less than $H_{max} - 14.9 / 2000 = H_{max} - 0.00745$.

Clearly eligible distributions that significantly deviate in entropy from the maximum value are very rare. However this result does not directly answer the question of how much confidence to have in the individual probabilities associated with distributions having maximum or almost maximum entropy. That is, can a probability distribution with close to maximum entropy assign probabilities that are significantly different from the probabilities of the maximum entropy distribution?

For instance, a 6-sided die having two faces with probabilities 1/12 and 1/4 respectively and four faces each having 1/6 probability has entropy 0.0436 less than the maximum of 1.7918 for a fair die. So for N=1000 independent trials and a significance level of $\alpha = 0.05$ such a distribution would differ from the maximum entropy value for a fair die by considerably more than 0.0047. However for N=100, $\Delta H = 9.49/200 = 0.047$, which is large enough to include such a distribution.

It should be remembered that maximum entropy works best where it is needed most - in situations involving a relatively large number of variables or values. In small sample situations it can produce obviously erroneous results.

Furthermore, how does the confidence in the probabilities determined by a maximum entropy solution depend upon the amount of under-specification of the situation that produced that solution? Surely a maximum entropy distribution that relies upon a great deal of ignorance about a situation offers less confidence about the probabilities determined than does a maximum entropy solution that is based upon a minimum of ignorance about the situation. Put another way, confidence about the maximum entropy distribution should be higher when conditional independencies are positively known than when they are merely provisionally assumed.

Amari [167, 169] takes up these issues in the context of differential geometry. Under-specification of information gives rise to a manifold of possible probability distributions. A Riemannian metric on these distributions early introduced by C. R. Rao [170] allows a very general approach to quantifying the distance between distributions. This development provides a very general approach to these problems of multiple possible distributions, but so far the results don't seem to directly apply to the issue of the confidence to be attached to the individual probabilities dictated by a maximum entropy distribution. Unfortunately Amari offers no numerical example to illustrate how these results might be applied to allow a confidence measure to be put upon the probabilities associated with distributions having maximum or close to maximum entropy.

Caticha [168] frames the question along the same lines as Jaynes: "Once one accepts that the maximum entropy distribution is to be preferred over all others, the question is to what extent are distributions with lower entropy supposed to be ruled out?" Using a parameterized family of distributions Caticha shows how this question can be rephrased as another maximum entropy problem, but he too offers no simple illustrative example of how his results can be applied to the question of how much confidence to have in any one probability value associated with the maximum entropy distribution.

What seems to be needed is a way to solve for the probabilities of specified outcomes in terms of entropies equal to or close to the maximum entropy. If 95% of the eligible probability distributions have entropy H no less than $H_{max} - \Delta H$, then what confidence limits are implied for the individual probabilities of those distributions?

6.9 Summary

In order to adequately represent and manipulate explicitly conditional statements such as "A given B" the familiar Boolean algebra of propositions or events must be extended to ordered pairs of such propositions or events. This is quite analogous to the requirement to extend integers to order pairs in order to adequately represent fractions and allow division. The resulting system of Boolean fractions includes the original propositions and also allows the non-trivial assignment of conditional probabilities to these Boolean fractions. Boolean fractions are truth functional in the sense that their truth status is completely determined by the truth or falsity of the two Boolean components of the fraction. But since there are two components, the truth status of a Boolean fraction has three possibilities – one when the condition (denominator) is false and two more when the denominator is true. Just as all integer fractions with a zero denominator are "undefined", so too are all Boolean fractions with a false condition undefined or "inapplicable". When the condition is true then the truth status of a Boolean fraction is determined by the truth of the numerator. The four extended operations (or, and, not, and given) on the Boolean fractions reduce to ordinary Boolean operations when the denominators are equivalent. Just as with integer fractions, the system of Boolean fractions has some new properties but loses others that are true in the Boolean algebra of propositions or events.

A conditional statement is not an implication or a deduction; it is rather a statement in a given context. Deduction of one conditional by another can still be defined in terms of the (extended) operations, as is often done in Boolean algebra. Due to the two components of a conditional there is a question of what is being implied when one conditional implies another. It turns out that several plausible implications between conditionals can be reduced to ordinary implications between the Boolean components of the two conditionals. The applicability, truth, non-falsity or inapplicability of one conditional can imply the corresponding property in the second conditional. Any two or more of these four elementary implications can be combined to form a more stringent implication. With respect to any one of these implications, a set of conditionals will generally imply a larger set, and it is now possible to compute the set of all deductions generated by some initial set of conditionals, as illustrated by three examples in this paper.

While computations can be done in principle, in practice the complexity of partial and uncertain conditional information precludes the possibility of solving for all possible probability distributions that satisfy the partial constraints. What is feasible and already successfully implemented in the program SPIRIT is to compute the distribution with maximum information entropy. However, the amount of confidence that can be associated with the probabilities assigned by this "most likely", maximum entropy distribution is still a somewhat open question.

> "Information theory provides a constructive criterion for setting up probability distributions on the basis of partial information, and leads to a type of statistical inference which is called the maximum-entropy estimate. It is the least biased estimate possible on the given information; i.e. it is maximally noncommittal with regard to missing information."
> E.T. Jaynes [180], Information Theory and Staistical Mechanics, 1957

Chapter 7

Fuzzy Sets, Time and Wholly-True Propositions

> A fuzzy set is a class of objects with a continuum of grade of membership. Such a set is characterized by a membership (characteristic) function which assigns to each object a grade of membership ranging between zero and one.
>
> L. A. Zadeh [58], "Fuzzy Sets", 1965

7.1 The Probability of Entailment --- The Fuzzy Connection

Assigning a probability to the proposition "q $\in$ (p)" would mean that the ideal (p) is to be a fuzzy set [58] because there would be some probability that the ideal subset (p) contains the proposition q. This would be to ask about the probability that q is always true if p is (given) true. Equivalently, $P(q \in (p)) = P(q \vee \sim p = 1)$. Perhaps this can even be expressed as $P(P(q|p) = 1)$.

Contrast this with the probability of q given p. The difference between $P(q \in (p))$ and $P(q|p)$ can be grasped more simply by considering the difference between P(q is always true) and P(q). P(q) is P(m(q)), the probability of the models in which q is true. But P(q is always true) asks about the probability that $q = 1$, that q is in the ideal set of theorems T.

One can hardly speak about "the models in which $p \equiv 1$ is true", because $p \equiv 1$ means that p is always true—true in all models. The issue is not how often p is true but rather how often p is always true! Clearly p can sometimes be "always true" only by there being some question (a probability) that the ideal of true propositions contains p. Here, the ideal of implicitly true propositions generated by the assumed axioms is variable. The various models that satisfy one or more of these variable axiom systems form a generally larger universe of models. From the extensional model (event) point of view, the universe Ω, of models that satisfy one set of axioms, A, has a probability in the larger universe of all possible models that satisfy at least one axiom system among a collection of axiomatic systems. In this context the set of wholly true propositions becomes a fuzzy set because there is some ambiguity about its membership.

So the probability of $P(p \equiv 1)$ may be different from $P(p)$. For example, consider a probabilistic game in which the rules change with certain probabilities that are more or less independent of the probabilities in the game, once the variable rules are temporarily fixed.

Competitive business must play by the legal rules, but these rules are subject to legislative change. A game of dealer's-choice poker has probabilities for the distribution of games that each dealer chooses as well as different probabilities within each different kind of poker game.

As mentioned before, one approach to an elaboration of these ideas would make the ideal of true propositions into a fuzzy set that may or may not contain a given proposition p, thus making it always true (or not). A less fuzzy way would postulate some sort of probability measure P_T, defined upon a σ-algebra $\mathcal{B}$ of ideals T of L including the zero ideal (0) and the one ideal (1), and satisfying the usual axioms of probability:

$$P_T(I) > 0 \text{ for all I in } \mathcal{B}, \tag{7.1.a}$$

$$P_T((1)) = 1, \tag{7.1.b}$$

$$P_T(I \vee J) = P_T(I) + P_T(J) \text{ if } I \wedge J = (0). \tag{7.1.c}$$

Note that $I \wedge J = (0)$ if and only if there is a proposition $p \in L$ such that $p \in I$ and $\sim p \in J$.

L. Zadeh [58, 98] has well demonstrated the usefulness of words that are better left imprecise, that are purposely imprecise in order to accurately express the degree of imprecision in the knowledge being communicated. When the word "tall" is applied to individuals of the human population, what is meant greatly depends on the context (subpopulation) of individuals about whom the statement is being made. A 6-ft. height man would be "tall" as a member of the whole human population but he would not be "tall" as a member of a professional basketball team. Even within the context of the NBA, the word "tall" is still imprecise because the set of players who are "tall" is "fuzzy". That is, there are different opinions about exactly when "tall" becomes "non-tall" for a NBA basketball player. The set of players for whom "tallness" is true is a fuzzy set because there is some uncertainty about its membership, an uncertainty that is not present for a "crisp" set, one for which the definition of membership determines whether or not any given element is a member.

The words "few", "some" and "several" demonstrate that even for a single individual in a given context, there may be no precise boundaries. In the context of people sitting in a living room, a "couple" is definitely 2, but "a few" appears to have a crisp lower boundary (namely 3) but a fuzzy upper boundary including at least 4 and 5 as possibilities. "Several" seems to be fuzzy on both ends, perhaps having lower boundary of 5, 6 or 7 and upper boundary from 8 through 11. Nevertheless, no one would say "several" and mean "3", nor mean 36 or 48. But context can change "few" to include 1000 if "few" refers to "a 1000 United Nations soldiers" sent to be observers of a war zone between two armies of 50,000.

7.2 The Fuzzy Connection

An important subcategory of fuzzy sets are those that arise when the logical situation embraces some variation of the laws (axioms, wholly true propositions) T upon which events are conditioned. The axioms determine an ideal $\mathcal{T}$, a deductively closed set of propositions that are wholly true - true in all models satisfying the axioms. However, if the set of axioms is variable, then a proposition may go from being wholly true to partially true or even false as the axioms change.

Corporations make profits within the laws of the country, but those laws can be changed making a formerly wholly true proposition no longer so. Similarly, in a game of "dealer choice poker" each player in turns gets to decide whether deuces will be "wild" (stand for any card in the 52 card deck). With deuces wild, poker hands of 5 Aces are possible and that hand wins in every game. But without deuces (or other cards) wild, the highest poker hands are the 4 royal, straight flushes (the highest 5 cards of any one of the 4 suits – spades, hearts, diamonds or clubs). Therefore, a royal straight flush will never lose when there are no wild cards, but this statement is false when there are wild cards in the games.

Similarly, the dice game of Craps displays this same kind of situation: An initial roll of two dice results either in five ways for the game to end (two winning and three losing) or the player rolls again. On the second (and any subsequent rolls) rolling a sum of 7 loses; rolling the same sum as the initial roll wins. Rolling any other sum does not apply. So, the player rolls again.

7.3 Time

What seems to be characteristic of these situations is a time sequence of decision moments, a sequential decision experiment in "time". Any logical understanding can be changed by subsequent events not accounted for in the axioms of that understanding. In general the ideal of wholly true propositions must be indexed by a "time".

Nevertheless, some propositions remain always true, others were true but not now, and some not true before, or now, will become true in the future. Other propositions will be wholly true some of the time.

Each proposition p can carry a time stamp t. This could be expressed as (p | time=t). That is "proposition p, given that the time = t". The conditional (p | time=t) simply requires that the time t be a P-measurable random variable on the collection of all "times" S. That is, there is a probability space $\mathcal{P} = (S, \mathcal{T}, P)$ where S is a set of all times t, $\mathcal{T}$ is a sigma-algebra of subsets of times, and P is a probability measure on $\mathcal{T}$. This fits well into the framework of Chapter 4, concerning operating on functions with variable domains and in particular, in the framework of conditional random variables.

7.4 Logical Models as Instants of Time

An even simpler formulation is possible by identifying an instance ω with an instant of time t. As previously developed, a conditional proposition can be represented as a partially defined P-measurable indicator function $p(\omega)$ on the set of all instances (models) Ω. Identifying instants of time t with instances ω allows time to be incorporated simply as a model, an instance. We can then speak of the probability that a fixed proposition p_0 is wholly true during a measurable subset T of all times S. The set of times t during which p is wholly true, that is, when $p_0(t) = 1$, is simply the inverse image of 1, namely, $p_0^{-1}(1)$.

Given a specific set of times T, there is a *conditional* probability $P(p_0|T)$ that the fixed proposition p_0 will be true among the times of T.

But if the set T varies, then there is a different probability P_T that $T \subseteq p_0^{-1}(1)$, that is, that during all times ω of T, $p_0(\omega) = 1$.

This probability is the probability of the ideal $(p_0^{-1}(1))$ generated by the set of times $p_0^{-1}(1)$ for which $p_0 = 1$. In symbols, this is $P_T((p_0^{-1}(1)))$. By distinguishing the two probability functions P and P_T there can be no confusion as to which probability function applies, and the double parentheses can also be eliminated. Thus $P_T(p_0^{-1}(1))$ is the probability of the ideal $((p_0^{-1}(1)))$ generated by the subset of times when p_0 is *wholly* true. [By contrast, $P(p_0^{-1}(1))$ is the probability of the subset of times (of S) when p_0 is true, which is the definition of $P(p_0)$.]

Thus for any proposition p, $P_T(p \in T)$, the probability that p is in the ideal T of always true propositions, is $P_T(p^{-1}(1))$. This makes the ideal T of always-true propositions a fuzzy set since there is only some probability that a proposition p is in it.

$T \subseteq p^{-1}(1)$ can also be formulated as follows: "if $t \in T$ then $t \in p^{-1}(1)$", which can be expressed as a conjunction of conditionals ($p(t) = 1 \mid t$) for all t, which is simply that $p(T) = \{1\}$, following the standard definition for the image of a set T acted upon by a function p.

We often hear that on some day t, something new has happened (is true) that was not true before time t. A proposition p can have $p(t) = 1$ for $t > t_0$ and $p(t) = 0$ for $t \leq t_0$. Or p might vary between 1 and 0 for some disconnected subset of times. The proposition p that the sun is visible in San Diego is true during the times after "sunrise" and before "sunset" when it is also not cloudy.

7.5 Fuzzy Boundaries

It should be noted that by the definition of a model ω (or time t), a proposition p is either true for ω or false for ω, either true at time t or false at time t. It cannot be ambiguous. Therefore, any ambiguity in the fuzzy boundary of a fuzzy set must be represented by varying the time (context) so as to include an element as a member of the set or not. For example, suppose that there is a 1/2 probability of my including "12" in "several". Then the situation could be represented as two equally likely sub-times t and s such that "several" includes 12 at time t, but not at time s.

More generally, using A. N. Kolmogorov's fundamental theorem [28] any consistent set of probabilistic constraints, such as a distribution function expressing the probability of a fuzzy set boundary can be expressed in terms of random variables on an appropriate sample space. Any individual time interval [t, s] can be subdivided into sub-instants of varying length for which a proposition p is true in some and false in others, or for which a random variable X has values within some Borel-measurable subset of times.

Some fuzziness is caused by inconsistencies in the data, data that nevertheless may have very useful information in a consistent subset of that data. Or it may simply be ambiguous. The latter kind of data processing will not be attempted here.

Having incorporated propositions whose truth depends upon time, and considering the eventual desire to incorporate multiple decision makers or agents whose knowledge at a time t will need to be expressed, a note of caution is appropriate:

7.6 Paradoxes of Self-Reference

If propositions carry a "time" stamp p_t then self-referential statements can lead to hidden inconsistencies such as with the "truthful judge", who tells the convicted man that he will be hung at high noon within a week, but that he will not know which day until the jailor tells him on the morning of the execution day. The man reasons that the truthful judge can't wait until the last day because the convict would know the night before, not on the day of the execution. So there are really only 6 days when the execution can be done if the judge is to be truthful.

But the same argument can then be applied to the 6th day, the 7th having been ruled out. And by repeating the argument he proved to himself that the truthful judge had no remaining days to schedule the hanging. Of course, one day during the week the jailor surprised the convict with the news of it being his execution day. But what was wrong with his reasoning?

This logical conundrum, which I learned as a student from Professor Karl Menger, is based on statements about knowledge at other times. It can be simplified down to just two possible hanging days: That the convict must not know before the day of his execution implies that the second day cannot be the hanging day. So it must be the first day. But then he knows that ahead of time. Or does he?

That the convict would be hung within a week is a statement about a future event. But the statement that the convict would not know the execution day until that morning is a meta-statement about prior knowledge applying to a future time.

Such self-referential statements can easily lead to inconsistencies as in the famous example: "The following statement is false. The previous statement is true." These are mutually inconsistent meta-statements.

There is also B. Russell's [50] famous self-referential example: "The set of all sets that do not contain themselves as a member" is non-existent because such a set would contain itself if it didn't but not contain itself if it did.

For these reasons it seems prudent here, at least for the time being, to confine temporal logic to changes over time in the truth of propositions or events that are not statements about the future knowledge of past events.

For example, the time evolution of quantum systems [99] satisfies this non-self referential condition even as it allows propositions about the history of the quantum system. L. Vanni & R. Laura have incorporated a time stamp t, on each proposition p, as an ordered pair (p;t), that "proposition p is true at time t".

J. Tyszkiewicz, A. Hoffmann, & Arthur Ramer [87] divide time into n discrete "states" and define a stochastic process with a transfer function mapping conditional propositions from one state to the next where they may change between being true, false or undefined.

7.7 Time and Quantum Logic

In "The Logic of Quantum Measurements" [100] using their innovative "contexts of histories" idea [99], L. Vanni & R. Laura describe a quantum system that starts with a property p_1 at a time t_1 and ends with a property p_2 at a future time t_2, all of which is determined by the "time evolution" (physical mechanics) of the system as given by its Hamiltonian operator H. The propositions are represented in the usual way as projectors in the Hilbert Space H of the quantum system. A property p_1 at time t_1 denoted $(p_1; t_1)$ is equivalent to a property p_2 at time t_2 denoted $(p_2;t_2)$ in case there is a unitary operator

$$U(t_1, t_2) = \exp(-iH(t_2 - t_1)2\pi/h) \qquad (7.7.a)$$

for which $(p_2; t_2) = U(p_1; t_1)U^{-1}$. These unitary translations establish an equivalence relation between properties at different times such that each equivalence class constitutes a possible time evolution of the quantum system.

As usual, at a given time t, a property $(p_1; t)$ with an associated vector subspace $(p_1; t)H$ *implies* a property $(p_2; t)$ with associated vector subspace $(p_2; t)H$ just in case

$$(p_1; t)\mathcal{H} \subset (p_2; t)\mathcal{H} \qquad (7.7.b)$$

This definition of implication is *extended* to properties existing at *different* times by translating both properties to a common time and seeing that the subspace inclusion holds.

The logical conjunction $(p_1; t_1) \wedge (p_2; t_2)$ of two property-time pairs is defined to be that property at a common time t whose subspace is the largest one contained in the subspaces of both $(p_1; t_1)$ and $(p_2; t_2)$ when translated to time t. The disjunction is defined similarly using the sup rather than the inf of the corresponding subspaces. The negation of a property-time pair (p;t) is defined as that property whose subspace is complementary to that of (p;t). The resulting algebra is a non-distributive, orthocomplemented lattice.

Laura & Vanni note that equivalent pairs, $(p_1; t_1)$ and $(p_2; t_2)$, are assigned the same probability in standard quantum mechanics because each can be transformed by quantum state operators ρ_1 and ρ_2 into operators whose traces are equal. If $(p_1; t_1)$ implies $(p_2; t_2)$ then the conditional probability $P((p_2; t_2) \mid (p_1; t_1)) = 1$.

Vanni & Laura restrict attention to logical conjunctions of projectors that commute when translated to a common time t. Such property-time pairs are called "contexts of histories". Others conjunctions are said to have no meaning in standard quantum mechanics nor probability frequencies.[16] The conjunctions of any finite number of these commuting property-time pairs generate atoms whose disjunctions form a distributive, orthocomplemented lattice.

Next, to the Hilbert Space representation of this quantum system, Vanni & Laura adjoin a macroscopic observable A with eigenvalues, a_i, associated with a measurement instrument. This allows statements of the form $(q_i; t_1) \wedge (a_i; t_2)$ where q_i is the value of some quantum ob-

---

[16] However, see Calabrese, Philip G. Toward a more natural expression of quantum logic with Boolean fractions. J. Philos. Logic 34 (2005), no. 4, 363--401. MR2181132 (2006h:81012) concerning simultaneously verifiable conditional events (p385), which exist in a larger algebra that includes all superpositions of events, events with incompatible conditions, and assigns them conditional probabilities.

servable Q at time t_1 and a_i is the value of the measuring instrument at a later time t_2.

This theoretically allow statements of the form "the quantum system had $Q = q_i$ before the measuring instrument indicated $A = a_i$". However, in the general case considered, the observable Q may be in a superposition of its eigenstates at the initial time t_1. This calls into question the very meaning of a property value of Q prior to the measurement. The Vanni & Laura attribute this ambiguity to the quantum entanglement of particles and designate the prior assignment of values as not completely objective, but merely "effective", an assignment dependent on the measurement, which it is.

However, this also reveals the inherent weakness of the current interpretation of a superposition of quantum events as being a mysterious simultaneous straddling of incompatible events, as though the proverbial quantum cat is both alive and dead. The alternate (DeBroglie-Bohm) interpretation postulates faster-than-light "guiding waves" to generate entanglement correlations while it allows local properties to maintain their values as actual even though measurement indeterminacy limits the precision of quantum value assignments and even initial conditions to "states" that are probability distributions.

7.8 Time Revisited

Time seems to have an arrow, a direction: past⇒present⇒future. However, this is not adequately conceived as a linear sequence of points any more than a real number line segment can be reduced to a set of points of zero length. In addition, as for the real line, at least all *time intervals* of positive "length" must be included to capture the *duration* or span of "times". That is how Alfred North Whitehead [50] framed time, and it satisfies the measurability requirements of probability theory. Also see Arthur N. Prior [40-41]

For purposes of calculating probabilities over "times", cyclic motions in space define relative time measurements of other space motions, any one of which defines a time duration event (t, s) relative to the observer, where t is the initial instant and s is the ending instant.

An "instant" t conceives all motion in space as "frozen in place" including the truth, falsity or inapplicability of any conditional proposition (p(t) | q(t)) at time t.

To capture the ordering of times, an adequately large, but finite interval of real number "instants of time" will suffice to form the Universe of all times, and its length can be normalized to unit length (1). Thus the real number unit interval [0,1] suffices as the universe Ω of a probability space with probability function P represented as "length" L.

A (Borel) sigma-algebra $\mathcal{B}$ is defined as all possible unions ($\cup$) of the intersection ($\cap$) of any *finite* number n of individual time intervals $[t_1, s_1], [t_2, s_2]... [t_n, s_n]$ of non-negative length $(s_i - t_i)$. Two disjoint members of $\mathcal{B}$ clearly have no times in common and so the length L of their union is the sum of their individual lengths, as required for a probability space.

In this way, the triple $\{[0,1], \mathcal{B}, L\}$ becomes a probability space $\mathcal{P}$ whose universe of "all times t" is the unit real number interval [0,1], and whose sigma-algebra $\mathcal{B}$ of subsets is the Borel algebra of subsets of the unit interval, and thirdly whose probability measure is the "length" or "measure" of those Borel subsets. This can all be accomplished with standard measure theory techniques.

Any number of intermediate times or "way points" may be imagined within any time interval [t, s] of positive length. As time ranges from t = 0 to t = 1, a conditional proposition $(a_t|b_t)$ can fluctuate from being false to being true to being inapplicable.

7.9 Indicator Functions

We may consider a conditional proposition (q|p) to be a partially defined indicator function $(q|p)_t \equiv (q_t \mid p_t)$, where q_t and p_t are propositions defined on the set [0,1] of all times t.

Alternately, we may consider each time t to be a partially defined indicator function that assigns to each conditional proposition (q|p) a value $t_{(q|p)}$ indicating its inapplicability (p is false), or its truth (1) or its falsity (0) at (or for) time t.

For some purposes we may want to represent both the conditional proposition (q|p) and the time t as (q|p, t). This would be "the conditional proposition 'q given p' at time t". This could represent a transition proposition to q given node p in a tree diagram having various possible nodes at a specified time t.

By specifying a finite set of times $t = t_1, t_2, ... t_n$, a proposition p_t defines a two-valued finite state stochastic process $\{p_1, p_2, ... p_n\}$ with probabilities that in general depend upon the context at all previous times. These can be extended to conditional propositions $\{(q_1|p_1), (q_2|p_2), \ldots (q_n|p_n)\}$.

Instead of propositions or conditional propositions, conditional (3-valued) random variables may be indexed by time as formulated by J. Tyszkiewicz, A. Hoffmann, & Arthur Ramer [87]

Chapter 8

A More Natural Expression of Quantum Logic with Boolean Fractions

> While we doubters have not shown so much self-confidence, nevertheless for all these years it has semed obvious to me – for the same reasons that it did to Einstein and Schrödinger – that the Copenhagen interpretation [of quantum mechanics] is a mass of contradictions and irrationality and that, while theoretical physics can of course continue to make progress in the mathematical details and computational techniques, there is no hope of any further progress in our basic understanding of Nature until this conceptual mess is cleared up.
> E.T. Jaynes [179], "Probability in Quantum Theory", 1990

8.1 Introduction

In this chapter the non-distributive system of Boolean fractions (a|b), where a and b are 2-valued propositions or events, is used to express uncertain conditional propositions and conditional events. These Boolean fractions, 'a if b' or 'a given b', ordered pairs of events, which did not exist for the founders of quantum logic, can better represent uncertain conditional information just as integer fractions can better represent partial distances on a number line. Since the indeterminacy of some pairs of quantum events is due to the mutual inconsistency of their experimental *conditions*, this algebra of conditionals can express indeterminacy. In fact, this system is able to express the crucial quantum concepts of orthogonality, simultaneous verifiability, compatibility, and the superposition of quantum events, all without resorting to Hilbert space. A conditional (a|b) is said to be "inapplicable" (or "undefined") in those instances or models for which b is false. Otherwise the conditional takes the truth-value of proposition a. Thus the system is technically 3-valued, but the 3rd value has nothing to do with a state of ignorance, nor to some half-truth. People already routinely put statements into three categories: true, false, or inapplicable. As such, this system applies to macroscopic as well as microscopic events. Two conditional propositions turn out to be simultaneously verifiable just in case the truth of one implies the applicability of the other. Furthermore, two conditional propositions (a|b) and (c|d) reside in a common Boolean sub-algebra of the non-distributive system of conditional propositions just in case b=d, their conditions are equivalent. Since all aspects of quantum mechanics can be represented with this near classical logic, there is no need to adopt Hilbert space logic as ordinary logic, just a need perhaps to adopt propositional fractions to do logic, just as we long ago adopted integer fractions to do arithmetic. The algebra of Boolean fractions is a natural, near-Boolean extension of Boolean algebra adequate to express quantum logic. While this explains one group of quantum anomalies, it nevertheless leaves no less mysterious the 'influence-at-a-distance', quantum entanglement phenomena. A quantum realist must still embrace non-local influences to hold that "hidden variables" are the measured properties of particles. But that seems easier than imaging wave-particle duality and instant collapse, as offered by proponents of the standard interpretation of quantum mechanics.

8.1.1 Controversies of Quantum Logic
The many questions and controversies (Putnam [103-104], Suppes[105-106], Wilce[108]) engendered by the measurements and interpretations of quantum mechanics (Planck[109], Einstein[110], de Broglie[111], Schrödinger[112], Dirac[113], von Neumann[114], Birkhoff[115]) can be divided into two categories. The first kind arises from the way measurements of one quantum variable like position can interfere with the conditions for precisely measuring another variable like velocity – the *indeterminacy* of some pairs of measurements whose experimental conditions are mutually inconsistent. (Heisenberg [116]; Bohm [117, p74]; Fine [118, p185])

The second kind of controversy focuses on the so-called "*measurement problem*", the apparent need in quantum mechanics for some type of space-transcending action or influence to explain quantum wave-particle duality as in the double slit experiments and especially the mysterious "particle entanglement" phenomena. (See de Broglie [111], Bell [119-121], Bohm [117], Goldstein [122-124].)

This chapter explains the logical principle of indeterminacy in terms of the *inapplicability* of statements, the third possible truth status of any statement besides *true* and *false*. This solution will clearly separate indeterminacy and superposition issues from issues regarding wave-particle duality and the measurement problem, which are different aspects of quantum mechanics. (See Wilce [108].)

8.1.2 Boolean Fractions

In that regard, the main purpose of this chapter is to show that a new, little known, *non-distributive,* 4-operation system of Boolean *fractions*, also called *conditional events*, is a more natural and much simpler algebraic context in which to express quantum mechanics.

That is, if one desires a simple, non-distributive system with which to adequately describe quantum propositions, one need look no further than to the system of Boolean fractions, conditional events, under operations of "and", "or", "not" and "given". This system of ordered pairs of propositions [63, 64] successfully finessed D. Lewis's triviality results [61] by providing a near-Boolean logical object (a|b) that is also able to non-trivially carry the conditional probability P(a|b). W. Rödder [157, 125] has successfully used the four operations in a very impressive interactive computer program called SPIRIT for quickly calculating Bayesian solutions using maximum entropy to assign values to unknown variables. A full theory of deduction for these uncertain conditionals [65, 69] is now available. These methods provide a natural way to extend numerical operations and averaging to real functions with variable domains ([70]; See Chapter 4.) This system of conditional events, which did not exist for von Neumann and Birkhoff, now appears to adequately and more intuitively describe quantum logical relationships of orthogonality, simultaneous verifiability, compatibility, and superposition.

8.1.3 Three-Valued Logic

In "Is logic empirical?" H. Putnam [103] claims that just as geometry needed revision when it became non-Euclidean so does 2-valued logic need revision in light of the requirements of quantum logic, that a 3-valued, non-distributive revision of ordinary logic is necessary because quantum logic requires one, in particular, that non-distributive superpositions of quantum events require such a non-Boolean, 3-valued logic and so that makes 3-valued logic the standard for doing logic in general [103].

Many objections [126, 156] have been raised against this program among them that i) 3-valued truth tables would muddy the clarity of logic, ii) Hilbert space logic hardly applies to events outside of quantum mechanics; iii) If 3-valued logic is necessary for quantum events, then in principle at what physical scale does one go from 3-valued logic to the ordinary 2-valued logic that we all use in the macro world? And a crucial question: iv) What is to be the interpretation of the 3rd truth-value - "unknown", "unknowable" or something else?

8.1.4 Conditional Event Logic
This chapter offers answers to these questions. It leaves 2-valued logic untouched, but subsumes it in a larger, 3-valued, non-distributive logic founded on pairs of 2-valued propositions. This new 3-valued logic of conditionals is quite applicable to macroscopic events, not just quantum events. So there is no problem answering the question of exactly at what physical scale one should transfer to the new logic; it applies to all physical scales. It includes ordinary Boolean logic; it does not undermine it.

Again, this 3-valued logic is not defined on single propositions or events, which remain 2-valued, but on ordered pairs of 2-valued propositions, Boolean fractions (a|b), "a given b", which include the original propositions as the set of fractions (a|Ω), where Ω is the universal event or the certain proposition.

Any 3-valued conditional proposition (a|b) is just an ordered pair of 2-valued propositions, and all operations on these 3-valued conditionals will reduce to operations on 2-valued propositions. So Putnam [103] is too strong in saying that 3-valued logic is "necessary" to represent quantum superpositions. It rather makes things easier to do and understand, something like how integer fractions make numerical calculations easier to do and understand.

8.1.5 Applicability - the 3rd Truth Status
The interpretation of the 3rd truth state or value for a conditional (a|b) is "inapplicable" whenever the condition b is false, whether or not a is true. If b is true, then (a|b) has the truth-value of proposition a. (a|b) is simply an ordered pair of 2-valued propositions or events which together form a 3-valued conditional proposition or conditional event.

So the 3rd truth-value here has nothing to do with incomplete knowledge or a state of ignorance. There are after all various 3-valued logics [48]. The one representing {true, false, unknown} is quite different from the one being promoted here. Here, conjunction, disjunction and negation operations are equivalent to the corresponding operations of the 3-valued logic of B. Sobocinski (See Rescher [46] p342). However, Sobocinski's logic has no 4th conditioning operation (|) and deduction is also defined differently.

The 3rd truth state here is not some half-true "middle" that muddies the logical clarity, but none other than our familiar "inapplicable". We are all used to declaring statements to be inapplicable, not just true or false, and the question of the applicability of a statement is a two-valued question.

The truth status of combinations of these 3-valued conditionals is fully determined by 3-valued truth tables, which are in turn completely defined in terms of the truth status of a pair of 2-valued propositions.

8.1.6 Conditional Probability – the Measure of Partial Truth

Contrary to the "material conditional" reduction of "a if b" into "a or else not b" which is used in 2-valued logic, an inapplicable conditional is not taken to be true. This is necessary also in order to measure the partial truth of a conditional (a|b) with the conditional probability P(a|b). The use of ordered pairs of propositions avoids the triviality results of D. Lewis [61] and allows (a|b) to carry the conditional probability of a given b, the numerical fraction P(a and b)/P(b). The probability of a proposition is defined as the probability of its extension, namely, the set of instances or models in which that proposition is true [63].

Conditional probability is also quite capable of consistently representing all of the formulas of quantum mechanics [127] as long as *changing conditions* are adequately represented, such as when the environmental configuration changes by opening up a second slit through which a particle can pass (See Feynman[128] pp 1-5; Koopman[129].

8.1.7 The Case for Non-Local Realism

Concerning the second category of controversy, in the "standard" Bohr interpretation of quantum mechanics, "wave-particles" extending into all space instantaneously "collapse" to one position when measured! There is a probability for each position. This interpretation avoids any further questions by claiming that these basic elements of physics (wave-particles) are otherwise amorphous and *unknowable* with any additional completeness. [130]

It is now more or less accepted that J. Bell's theorem [119], [131], [132] proved that any theory of "hidden variables" (properties associated with individual quantum particles that account for the results of quantum measurements) has to admit some kind of "non-local" (faster-than-light-speed) influence. Rather than admit such "spooky" action at a distance, the "standard" interpretation denies that the old words have any meaning in the quantum domain.

However, some type of non-locality has been a part of our understanding of physics at least since Newton, who postulated instantaneous attraction between distant bodies. According to Einstein's relativity theory, a massive body bends the space near and to some extent, far away from itself. Standard quantum mechanics claims that quantum particle-waves extending into all space instantaneously collapse to one position when measured. Lately, vibrating "strings" stretch far and wide at once tying distant places together.

The so-called measurement problem (how to understand quantum wave-particle duality and entanglement phenomena) does not go away when we go from a bivalent to a 3-valued formal logic that explicates indeterminacy. These are two different phenomena. Putnam's suggestion therefore amounts to trying to sweep both philosophical difficulties of interpretation of quantum mechanics under one 3-valued rug. The so-called Copenhagen interpretation of quantum events offers this cloud as the optimal perspective.

But since some kind of "influence at a distance" seems necessary in physical theories, perhaps it is time to simply accept the notion that space separation is not absolute, that some kind of unknown wave energy transcends space, potentially allowing events in distant positions to influence one another instantaneously, or very quickly. Quantum entanglement statistics is the pragmatic reason for this interpretation, but it also applies to the double slit, wave interference phenomena associated with individual particles.

After all, the universe, including space, is first of all a unified whole, not merely an aggregation of distance-related parts. Space is not pure emptiness, a geometric concept of the relatedness of points, but rather, a qualification of the totality of the cosmos. So local events might have distant influences via some 'space-transcending energy reaction of the universe'. There does not have to be anything truly metaphysical involved here, just a belief in the physical unity of the universe – some previously unknown kind of space transcending "pre-energy" obeying quantifiable physical laws.

For the price of this new cosmological framework, one gets to keep the realism of having actual particle properties associated with quantum events.

8.1.8 The Power of Ordered Pairs

It is not obvious that pairs of integers (ordinary integer fractions) would be so much better at representing distances on a number line than single integers. Nor is it obvious that uncertain conditional information is so much better represented and combined by having each proposition (or event) carry along its own condition, its domain of application. These Boolean fractions do for propositions & events what numerical fractions do for integer arithmetic; they dramatically expand what can be accurately represented.

This single change extends Boolean logic enough to allow one to express the logic of quantum measurements including the crucial phenomena of orthogonality (a kind of disjointness for conditional events), simultaneous verifiability, simultaneous falsifiability, compatibility and the superposition of quantum propositions, all without resorting to Hilbert space.

Because G. Boole was unsuccessful in adding a worthy division operation to his three other operations, Kolmogorov [133] was unable to explicitly incorporate changing conditions into his celebrated axiomatization of probability theory. Therefore was contextualism unable to derive quantum statistics (Khrennikov [137, p3]; [134-136]) and quantum logic turned to "non-contextuality" – events that have the same probability no matter what the initial context (Pitman [138] p5; [139]). Quantum measurements so disturb the whole system that the state (context) is always changing. The concept of conditional events (Boolean fractions) to describe conditional statements finally appeared in the 1960's, but by that time the orthodox quantum formulation was well established and spectacularly successful at predicting the statistics of quantum experiments. It was just as spectacularly unsuccessful at illuminating anyone's physical intuition about what was really happening. The strange quantum interference phenomena suggested to Bohr that no such intuitive or "hidden variable" interpretation is possible for quantum mechanics – the so-called Copenhagen interpretation. Since then people have been regularly revisiting the quantum formulation (Fuchs [140]) looking for another way to understand quantum mechanics. Well, they need look no further than to this system of Boolean fractions for such a natural and intuitive formulation of quantum logic.

8.1.9 Preview
Section 2 gives some background such as G. Schay's algebra (8.2.1) and the basics of the author's version of conditional event algebra (8.2.2-8.2.3, 8.2.10), and other preliminary results to be used to state and prove the main results of this chapter in Section 8.3. This includes necessary and sufficient conditions for distributivity (8.2.4-8.2.5), a die-throwing example of non-monotonic disjunction (8.2.11), some superposition formulas for conditionals (8.2.12), simultaneous physical measurements (8.2.14), orthocomplementation (8.2.15-8.2.16), orthogonality of conditionals (8.2.17-8.2.19), and complementation for conditionals (8.2.20). For the benefit of those pursuing different approaches, this section and Section 3 also contain theorems relating the logic of conditional events to other quantum logical concepts developed in the literature such as the additive law of probability (8.2.13). Section 8.3.1 is totally new. It defines simultaneous verifiability of conditionals (8.3.1), and intuitively characterizes it (8.3.2-8.3.3) thereby clarifying this concept. Simultaneous falsifiability of conditionals (8.3.4-8.3.5) does not automatically follow from simultaneous verifiability, but together they are equivalent to the two conditionals having equivalent conditions (8.3.6). These are the fully compatible conditionals. Section 8.3.7 proves that two conditionals (a|b) and (c|d) are in a common Boolean sub-algebra if and only if they have non-impossible, equivalent conditions. The corollary (8.3.8) is that conditionals are compatible if and only if they reside in a common Boolean sub-algebra of the non-Boolean algebra of conditionals. Relative negation of conditionals is unique (8.3.9). Conditionals satisfy most or all of the requirements of other quantum concepts in the literature including orthoalegbras (8.3.10-8.3.11) and orthocomplementation (8.3.12). New verifiability distinctions are possible with conditionals (8.3.18-8.3.19). Section 8.3.20 explores the correspondence between the subspaces of the algebra of conditionals and the closed linear spaces of Hilbert Space.

8.2 Boolean Fractions and Quantum Mechanics
Concerning the philosophical basis for insisting on ordered pairs of propositions for representing information, I wish to state categorically that all information has explicit or implicit assumptions that must be represented in any adequate logic of information. Any so-called material fact is based upon the assumption of the validity of the human sensory system and related systems. This is assumed context, condi-

tions for any and every material fact. Likewise every supposedly clear idea is founded upon assumptions of the validity of the human intellectual apparatus and conceptual framework. "I think" assumes that thought is real based upon nothing but personal human experience of what we all recognize as "thought". We assume that humans have common physical and mental experiences that can be mutually identified without further justification. But these constitute assumed conditions. And depending on the specific facts or ideas these conditions can easily vary.

So the overall premise of this chapter is that the basic unit of information is a pair, not a single event or proposition. Every proposition or event has implicit assumptions, context, a conceptual framework based on definitions that divide the totality of the universe of discourse. The basic unit of information must include a supposition (or condition) in the context of which another proposition or event is being identified.

Just as integer fractions dramatically improve the ability of integers to represent and operate on distances in space, so too do propositional fractions dramatically improve the ability of propositions to represent and operate on uncertain conditional information in a situation space of variables taking on various possible values.

8.2.1 The Conditional Event Algebras of Schay[17]

By 1968, G. Schay [60] clearly had been developing his "algebra of conditional events" for quite some time before he published his one 10-page article. In it, he quickly defined a conditional event as a 3-valued indicator function as done here but without insisting that it be measurable. However his notation makes it clear that he was thinking of these indicator functions as defined on some kind of probability sample space Ω. He defined two disjunction operations and two conjunction operations on pairs of these indicator functions as follows:

$$((A|B) \cap_s (C|D))(\omega) = \min\{(A|B)(\omega), (C|D)(\omega)\}, \text{ with domain } B \cap D,$$
(8.2.1)

---

[17] I wish to thank I. R. Goodman for discovering Schay's 1968 paper and bringing it to my attention.

$$((A|B) \cup_S (C|D))(\omega) = \max\{(A|B)(\omega), (C|D)(\omega)\}, \text{ with domain } B \cup D, \tag{8.2.2}$$

$$((A|B) \wedge_S (C|D))(\omega) = \min\{(A|B)(\omega), (C|D)(\omega)\}, \text{ with domain } B \cup D, \tag{8.2.3}$$

$$((A|B) \vee_S (C|D))(\omega) = \max\{(A|B)(\omega), (C|D)(\omega)\}, \text{ with domain } B \cap D, \tag{8.2.4}$$

and negation ($\sim$) on these indicator functions as

$$(\sim(A|B))(\omega) = (A'|B)(\omega), \tag{8.2.5}$$

where A' is the negation or complement of A. (Schay used the over bar notation, $^-$, for the complement of A.) An "s" subscript is attached to Schay's operations to distinguish them from the operations to be recommended here.

Almost as an after thought, and without stating it is as a definition, Schay inserted in the first sentence of his Theorem 1, the equivalence relation for pairs of propositions:

$$(A|B) = (C|D) \text{ if and only if } B=D \text{ and } A \cap B = C \cap D \tag{8.2.6}$$

He then pointed out (Theorem 1) that these operations on indicator functions can be expressed as operations on ordered pairs of propositions (conditional events) as follows:

$$(A|B) \cap_S (C|D) = (A \cap C | B \cap D), \tag{8.2.7}$$

$$(A|B) \cup_S (C|D) = ([(A \cup C) \cap B \cap D]$$
$$\cup (A \cap B \cap D') \cup (B' \cap C \cap D) | B \cup D) \tag{8.2.8}$$

$$(A|B) \wedge_S (C|D) = ((A \cap B \cap C \cap D)$$
$$\cup (A \cap B \cap D') \cup (B' \cap C \cap D) | B \cup D) \tag{8.2.9}$$

$$(A|B) \vee_S (C|D) = (A \cup C | B \cap D) \tag{8.2.10}$$

$$\sim(A|B) = (A'|B) \tag{8.2.11}$$

A typographic error in Schay's negation equation in Theorem 1 has mistakenly rendered it as $\sim(A|D) = (A'|B)$. In addition, the operation $\cup_S$ can obviously be simplified to

$$(A|B) \cup_S (C|D) = ((A \cap B) \cup (C \cap D) | B \cup D) \tag{8.2.12}$$

Concerning the second and third operations listed together with negation, Schay stated, 'As for the intuitive meaning of the operations, it seems that $\cup_s$, $\wedge_s$ and $\sim$ correspond to the usual meaning of the words "or", "and" and "not".'

In fact, these are the very same, independently discovered operations on conditionals called "quasi" by Adams [94], [2], [53, p164]. They were again independently discovered by Calabrese [63] as part of a richer system with a fourth, iterated conditioning operation.

Later in his article (p343), Schay noted that these operations give meaning to expressions like P((A|B) $\cup_s$ (C|D)), but that "such probabilities have strange properties". "It is simple", he said, "to construct examples such that P(x $\cup_s$ y) < P(x), or P(x $\wedge_s$ y) > 0 although P(y) = 0, etc." These are the same kind of non-monotonicities that prompted Adams to call these operations "quasi".

Schay also said that the usual rule P(A|B)P(B) = P(A $\cap$ B) can be dropped to handle special situations in which a change in condition "changes the results in an unusual way." His example was a 3-way election in which the probability of each candidate winning is 1/3, but if one candidate drops out, then one of the other two candidates gets all of his votes. Schay then said, "It may along these lines be possible to incorporate the probabilities of quantum mechanics in our theory".

However, such probability examples as the vote problem above can be better handled by ordinary probability theory by defining more events to represent the situation. For instance, by defining the events "a", "b" and "c" of having each candidate in the race, the fact that A inherits all of C's votes can be expressed as P(A | abc) = 1/3 and P(A | abc$'$) = 2/3, without violating standard conditional probability theory. So P(A) = (1/3)P(abc) + (2/3)P(abc$'$). Schay ended his paper suggesting connections be explored between his theory and quantum logic and quantum probability. A few years later in 1976, Hardegree [163] was investigating "the conditional in quantum logic".

While ignoring the system with operations $\{\cup_s, \wedge_s, \sim\}$, which he said seems to correspond to the usual meanings of the words "or", "and" and "not", Schay focuses instead on two other combinations, one with operations $\cap_s$, $\cup_s$ and $\sim$, and the other with operations $\wedge_s$, $\vee_s$, and $\sim$. His Theorem 2 (p336) states that both of these systems of conditional events form distributive lattices. Schay probably had quantum mechanics in mind when he defined his two systems since the

Hilbert space formulation of quantum mechanics starts with a (nondistributive) lattice. If so, the desire to form a lattice may have been misguided because his systems turned out to be distributive and so not applicable to quantum mechanics.

In an effort to extend the range of his operations to mappings such as (B | A ∪ B) → (0|A) and (B | A ∪ B) → (B|B), Schay defined four more operations. However, he did not define an iterated conditioning operation, which could have accomplished something like this in a less arbitrary way. For instance, using the iterated operation ((A|B) | C) = (A | B ∩ C), which works for conditionals as well as simple Boolean propositions or events, and assuming A and B are disjoint (orthogonal) this mapping can be obtained by conditioning (B | A ∪ B) by A and conjoining the result by (B'|A). (In this example, the order of these two steps can be reversed with the same result.)

$$((B | A \cup B) | A) \cap (B'|A) = (B|A) \cap (B'|A) = (0|A) \qquad (8.2.13)$$

In this regard Tyszkiewicz [86] has shown that all alternative operations on conditionals offered by various authors can be generated by composition of these four operations.

Schay's main result (Theorem 5) is a generalization of Stone's well-known theorem [54] characterizing any Boolean algebra as an algebra of subsets under union, intersection and complement.

8.2.2 Conditional Event Algebra

The conditional event algebra introduced by Calabrese in 1987 has been thoroughly developed ([63-65], [67]) including deduction of uncertain conditionals [69, 71]. It is completely consistent with standard conditional probability theory but extends it to 3-valued Boolean fractions (conditional events), which are "undefined" in case their condition is false.

Let ($\mathcal{B}|\mathcal{B}$) denote the set of ordered pairs, {(a|b): a, b in $\mathcal{B}$}, called the set of *conditionals*, "a given b", of $\mathcal{B}$. The proposition or event "b" is called the *condition, premise* or *antecedent* and the proposition or event "a" is called the *consequent* or *conclusion*. Define two conditionals to be equivalent (=) as in equation (8.2.6): (a|b) = (c|d) if and only if b=d and ab = cd.

The algebra of Boolean fractions provides four operations for "or", "and", "not" and also "given" (conditioning) to handle compound conditional forms such as "((a|b) | (c|d))".

Relative Negation The relative negation of "a given b" is the "negation of a, given b". That is,
$$(a|b)' = (a'|b), \qquad (8.2.14)$$
and the latter has probability 1 - P(a|b).

Disjunction Concerning disjunction, "if b then a, or if d then c" means "if either conditional is applicable then at least one is true". That is,
$$(a|b) \vee (c|d) = (ab \vee cd) | (b \vee d) \qquad (8.2.15)$$

Conjunction Concerning conjunction, "if b then a and if d then c" means "if either conditional is applicable then one is true while the other is not false". That is,
$$(a|b) \wedge (c|d) = [ab(c \vee d') \vee (a \vee b')cd] | (b \vee d) \qquad (8.2.16)$$
$$= (abd' \vee abcd \vee b'cd | b \vee d), \qquad (8.2.17)$$

which also means "if either conditional is applicable then either they are both true or else one is true while the other is inapplicable."

Iterated Conditioning A conditional (c|d) may itself be a condition for another proposition or conditional proposition; a fraction can be in the denominator.
$$((a|b) | (c|d)) = (a | b \wedge (c|d)) \qquad (8.2.18)$$
$$= (a | b (c \vee d')) \qquad (8.2.19)$$

The order of preference of the operations is negation (') before conjunction ($\wedge$), before disjunction ($\vee$), before conditioning (|).

The algebra ($\mathcal{B}|\mathcal{B}$) of conditionals includes the original Boolean algebra $\mathcal{B}$ as those conditionals (a|Ω), where Ω is the universal event, and "a" is any member of $\mathcal{B}$. In logical notation these are the conditionals (a|1) whose condition is certain. Analogously, these are like the integer fractions whose denominators are 1. Fixing the condition b also yields a Boolean algebra $(\mathcal{B}|b) = \{(a|b): \text{all } a \in \mathcal{B}\}$.

Among the properties retained by the algebra of conditionals are the two familiar de Morgan formulas by which conjunction can be expressed in terms of disjunction and negation, or disjunction can be expressed in terms of negation and conjunction.

8.2.3 Properties of Conditional Event Algebra

The properties of this conditional event algebra have been listed in [63] and completely characterized in [69]. The disjunction and conjunction operations are associative, commutative and idempotent. The zero conditional (0|1) is an absolute zero with respect to conjunction in the sense that (a|b) ∧ (0|1) = (0|1), but (a|b) ∨ (0|1) = ab rather than (a|b). Similarly, (1|1) is an absolute unity element with respect to disjunction in the sense that (a|b) ∨ (1|1) = (1|1), but (a|b) ∧ (1|1) = a ∨ b' rather than (a|b). However, any conditional (a|b) has a relative complement (a'|b) such that (a|b) ∨ (a'|b) = (1|b) and (a|b) ∧ (a'|b) = (0|b). Negation also satisfies the law of double negation: ((a|b)')' = (a|b).

In Boolean algebra the equation B ∧ A = B ∧ (A if B) always holds. This also holds with conditionals: (a|b) ∧ (c|d) = (c|d) ∧ [(a|b) | (c|d)] by expanding the right hand side using the operations on conditionals. Concerning distributivity there is the following result.

8.2.4 Theorem on Distributivity

(a|b) ∧ [(c|d) ∨ (e|f)] = (a|b)(c|d) ∨ (a|b)(e|f) if and only if $(ab)(e'f) \le d$ and $(ab)(c'd) \le f$. That is, conjunction distributes over disjunction just in case the truth of the outside conditional and the falsity of one of the inside conditionals implies the other inside conditional is applicable.

Proof of Theorem 8.2.4 [(c|d) ∨ (e|f)] = (cd ∨ ef | d ∨ f). Conjoining this with (a|b) yields (abd'f' ∨ (b' ∨ ab)(cd ∨ ef) | b ∨ d ∨ f). On the other hand by expanding and collecting terms it follows that (a|b)(c|d) ∨ (a|b)(e|f) = (abd'f' ∨ abd'f ∨ abdf' ∨ (b' ∨ ab)(cd ∨ ef) | b ∨ d ∨ f).

These two conditionals will be equal if and only if abd'f ∨ abdf' ≤ abd'f' ∨ (b' ∨ ab)(cd ∨ ef). Since (abd'f') is disjoint from (abd'f ∨ abdf'), this is equivalent to

$$abd'f \vee abdf' \leq (b' \vee ab)(cd \vee ef),$$

and since $= b'(cd \vee ef)$ is disjoint from $(abd'f \vee abdf')$, the inequality is equivalent to

$$abd'f \vee abdf' \leq abcd \vee abef.$$

Since $abd'f$ is disjoint from $abcd$, therefore $abd'f \leq abef$. Similarly, $abdf' \leq abcd$. On the other hand, if these latter two inequalities hold then the one above obviously holds. Finally, $abd'f \leq abef$ is equivalent to $abd'fe' = 0$, which becomes $(ab)(e'f) \leq d$, one of the two inequalities to be proved. Similarly $abdf' \leq abcd$ is equivalent to $(ab)(c'd)f' = 0$, which becomes $(ab)(c'd) \leq f$, the other inequality to be proved. That completes the proof of the theorem.

8.2.5 Corollary on Distributivity
Distributivity of disjunction over conjunction. $(a|b) \vee [(c|d) \wedge (e|f)] = [(a|b) \vee (c|d)] \wedge [(a|b) \vee (e|f)]$ if and only if $(a'b)(ef) \leq d$ and $(a'b)(cd) \leq f$. That is, disjunction distributes over conjunction if and only if whenever the outside conditional is false and one inside conditional is true then the other inside conditional is applicable.

Proof of Corollary 8.2.5 Using the de Morgan formula,

$$\begin{aligned}(a|b) \vee [(c|d) \wedge (e|f)] &= \{(a|b)' \wedge [(c|d) \wedge (e|f)]'\}' \\ &= \{(a'|b) \wedge [(c|d)' \vee (e|f)']\}' \\ &= \{(a'|b) \wedge [(c'|d) \vee (e'|f)]\}' \\ &= \{[(a'|b) \wedge (c'|d)] \vee [(a'|b) \wedge (e'|f)]\}'\end{aligned}$$

where by the theorem the last equality is true if and only if $(a'b)(ef) \leq d$ and $(a'b)(cd) \leq f$. But $\{[(a'|b) \wedge (c'|d)] \vee [(a'|b) \wedge (e'|f)]\}'$

$$\begin{aligned} &= [(a'|b) \wedge (c'|d)]' \wedge [(a'|b) \wedge (e'|f)]' \\ &= [(a'|b)' \vee (c'|d)'] \wedge [(a'|b)' \vee (e'|f)'] \\ &= [(a|b) \vee (c|d)] \wedge [(a|b) \vee (e|f)], \end{aligned}$$

and that completes the proof of the corollary.

8.2.6 Three-Valued Logic

It was B. Definetti [90] who first stated the three-valued nature of conditionals, that conditionals (Boolean fractions) can be represented by 3-valued measurable indicator functions. (Also see [63], p234.) Suppose that two measurable Boolean propositions "a" and "b" are represented by their associated measurable indicator functions:

$$a(\omega) = \begin{cases} 1, & \text{if a is true in state } \omega, \\ 0, & \text{if a is false in state } \omega, \end{cases} \text{ and}$$

$$b(\omega) = \begin{cases} 1, & \text{if b is true in state } \omega, \\ 0, & \text{if b is false in state } \omega. \end{cases}$$

In this formulation a Boolean fraction or conditional event is an indicator function (a|b) given by:

$$(a|b)(\omega) = \begin{cases} 1, & \text{if a and b are true in } \omega, \\ 0, & \text{if a is false and b is true in } \omega, \\ \text{Undefined}, & \text{if b is false in } \omega. \end{cases}$$

Each operation on conditionals easily generates a 3-valued truth table by setting a, b, c, and d equal to 1 or 0 in equations (8.2.14), (8.2.15), (8.2.17) and (8.2.19). The converse is also true. See [68, p7]. The four 3-valued truth tables for conditionals are listed in the following table:

	AND			OR			GIVEN			NOT
	T	F	U	T	F	U	T	F	U	
T	T	F	T	T	T	T	T	U	T	F
F	F	F	F	T	F	F	F	U	F	T
U	T	F	U	T	F	U	U	U	U	U

Table 8.1. Three-Valued Truth Tables for Conditional Propositions

8.2.7 Three-Valued (Conditional) Computations

While providing a clear and elementary account of the mathematics of the quantum formalism A. Landé [141, p436] said that "an artificial difference between the classical and quantum realms has been established by those who think that macroscopic physics can be understood in terms of ordinary logic, but that atomic physics requires the introduction of a *three-valued logic*, true, false and undetermined." "We do not deny", he says, "anyone the pleasure of this mental gymnastics. But why only in case of the quantum realm when this logic fits (or does not fit) ordinary chance situations as well!"

My response to this rhetorical challenge is to assert that mathematicians and physicists have in fact for a long time been ignoring the need for such a 3-valued logic in ordinary chance situations!

To demonstrate this little recognized fact, consider the following game of chance. An ordinary six-sided die with faces numbered from 1 through 6 is rolled once. Someone makes the following statement:

> If the roll is an even number then it will be a "2" or if
> the roll is "less than 5" then it will be "less than 4".

What is the probability of winning this bet? More importantly why is there no standard way of calculating this probability in terms of the component probabilities and conditional probabilities? Why is there no standard way of operating with the conditional events? Some people even claim erroneously that such compound conditional statements have no meaning, but there are all sorts of examples of these kinds of statements in natural language. (One need only say something like "if the store has soda then buy me some soda, and if the store has beer then buy me some beer" to have an example of a compound conditional English sentence that makes perfect sense.) See for instance [66, pp192-203] for several practical problems completely solved using this conditional event algebra.

The answer to this die problem above can be determined by brute force examination of each possible outcome from 1 to 6 to see if it first satisfies the condition and then whether it satisfies the consequent. In this way the outcomes 1, 2, 3, 4 and 6 are identified as satisfying one or both of the conditions of being "even" or "less than 5". Of these five outcomes, 4 or 6 will lose the bet while 1, 2, or 3 will win. So the probability of winning is 3/5. If outcome 5 turns up then the bet is neither won nor lost. This is the inapplicable case where the 3rd truth-value comes in.

But this probability could also have been calculated using the 3-valued operations on conditionals in the system of Boolean fractions as follows:

$$(2 \mid \text{even}) \vee (<4 \mid <5)$$
$$= (\{(2) \cap \text{even}\} \cup \{(<4) \cap (<5)\}) \mid (\text{even} \cup (<5))$$
$$= (\{2,1,2,3\} \mid \{1,2,3,4,6\})$$
$$= (\{1,2,3\} \mid \{1,2,3,4,6\}),$$

which has conditional probability 3/5. In more complicated situations the answer can be calculated using the general formula [65, p84]

$$P((a|b) \vee (c|d)) = P(a|b)P(b| b \vee d) + P(c|d)P(d| b \vee d)$$
$$- P(abcd \mid bd)P(bd \mid b \vee d)$$

$$= (1/3)(3/5) + (3/4)(4/5) - (1/2)(2/5) = 3/5$$

where a = (2), b = (even), c = (<4) and d = (<5), and juxtaposition has replaced conjunction ∧.

Notice that this "superposition" (disjunction) of two conditionals (a|b) and (c|d) that individually have conditional probabilities 1/3 and 3/4 respectively, has a combined probability value of 3/5, which is *between* 1/3 and 3/4, not above them both as is always true for Boolean propositions.

Taken in the abstract this feature of the algebra of conditionals seems unintuitive and hard to swallow. But it clearly happens in a simple die throw when two propositions with different conditions are disjoined or conjoined. The reason for the possibility of non-monotonicity when conditionals are combined is due to the possibilities of an expanded context (condition).

For instance, were someone to bet "if the die role is even then it will be a 2, 4, or 6", they would win with probability 1. But suppose the person said instead, "If the die role is even then it will be a 2, 4, or 6, or if the roll is odd then it will be a 5." Here, since the context (condition) has expended to all 6 outcomes, and because only 4 of them are winning outcomes, the probability of winning has gone down from 1 to 2/3 even though the person used "or" between the two clauses.

This is simply the non-monotonic way conditionals with different conditions operate. It is also easy to construct such simple examples in which $P((a|b) \wedge (c|d)) > P(a|b)$.

8.2.8 Disjunction and Conjunction Superposition Formulas

As stated in [67, p1682], without further qualification, for any conditionals (a|b) & (c|d),

$$(a|b) \lor (c|d) = (a|b)(b \mid b \lor d) \lor (c|d)(d \mid b \lor d).$$

This can easily be verified by applying the operations on conditionals to see that the two sides are equal. It is also easy to verify that

$$(a|b) \lor (c|d) = (a|b)(bd' \mid b \lor d) \lor (c|d)(b'd \mid b \lor d) \lor ((a \lor c)bd \mid b \lor d),$$
$$P((a|b) \lor (c|d)) = P(a|bd')P(bd' \mid b \lor d) + P(c \mid b'd)P(b'd \mid b \lor d)$$
$$+ P((a \lor c)bd \mid b \lor d)$$

Similarly,

$$(a|b) \land (c|d) = (a|b)(bd'|\ b \lor d) \lor (c|d)(b'd|\ b \lor d) \lor ((a \land c)bd \mid b \lor d),$$
$$P((a|b) \land (c|d)) = P(a|bd')P(bd' \mid b \lor d) + P(c|b'd)P(b'd \mid b \lor d)$$
$$+ P((a \land c)bd \mid b \lor d)$$

Notice also from Equation 8.2.16 that if (a|b) and (c|d) are two conditionals for which the truth of one implies the non-falsity of the other, that is, if $(ab)(c'd) = 0 = (a'b)(cd)$, or equivalently when (a|bd) = (c|bd), such as when b and d are just disjoint, then

$$(a|b) \lor (c|d) = (a|b) \land (c|d),$$

(and conversely) and so their probabilities are equal. In this case the last term in both of the above probability formulas becomes P(abcd | b ∨ d). If b and d are disjoint then the probability of both (a|b) ∨ (c|d) and (a|b) ∧ (c|d) is P(a|b)P(b | b ∨ d) + P(c|d)P(d | b ∨ d). This is a weighted average of P(a|b) and P(c|d) with weights P(b | b ∨ d) and P(d | b ∨ d) respectively expressing the relative probabilities of b and of d given that either is true. If also a=c, then

$$(a \mid b \lor d) = (a|b)(b \mid b \lor d) \lor (a|d)(d \mid b \lor d),$$
$$P(a \mid b \lor d) = P(a|b)P(b \mid b \lor d) + P(a|d)P(d \mid b \lor d).$$

Recalling the quantum formalism, note how [(b | b ∨ d), (d | b ∨ d)] is a unit vector representing a "state", and [(a|b), (a|d)] is some kind of projection (conditioning) of "a" onto b and onto d. In general this is just

$$P(a \mid u) = \Sigma_i P(a \mid u_i) P(u_i \mid u)$$

Concerning under what circumstances the conditional probability function is additive there is the following result:

8.2.9 Theorem on the Additive Law
$P((A|C_1) \vee (B|C_2)) = P(A|C_1) + P(B|C_2)$ if and only if one of the following is true:

1) $P(A|C_1) = 0 = P(B|C_2)$
2) $P(A|C_1) = 0$ and $C_1 \leq C_2$
3) $P(B|C_2) = 0$ and $C_2 \leq C_1$
4) $C_1 = C_2$ a.s. and $P(A \wedge B \mid C_1) = 0$

Proof of Theorem 8.2.9 In general $P((A|C_1) \vee (B|C_2)) = P(A|C_1)P(C_1 \mid C_1 \vee C_2) + P(B|C_2)P(C_2 \mid C_1 \vee C_2) - P(AC_1 BC_2 \mid C_1 \vee C_2)$. If this equals $P(A|C_1) + P(B|C_2)$, then rearranging terms yields

$$P(A|C_1)[1 - P(C_1 \mid C_1 \vee C_2)] + P(B|C_2)[1 - P(C_2 \mid C_1 \vee C_2)]$$
$$= - P(AC_1 BC_2 \mid C_1 \vee C_2).$$

Since all terms on the left hand side are non-negative, but the right hand side is non-positive, all terms must be zero. So $P(A|C_1)[1 - P(C_1 \mid C_1 \vee C_2)] = 0$, $P(B|C_2)[1 - P(C_2 \mid C_1 \vee C_2)] = 0$, and $P(AC_1 BC_2 \mid C_1 \vee C_2) = 0$. From these 3 equations the result easily follows by cases. That completes the proof of the theorem.

Thus only when the conditions C_1 and C_2 are equal is there a non-trivial realization of the sum formula for disjoint events A and B in the context of the common condition C_1.

P. Busch [142, p2] noted that when ψ is a proper superposition of eigenstates of a quantum observable B, then B is indeterminate in the state, ψ. That is, B does not have a definite value. This corresponds to the Hilbert space projection operator T for B satisfying neither $T\psi = \psi$ nor $T\psi = 0$. Unlike Boolean algebra, the system of Boolean fractions naturally includes conditional propositions B whose value $B(\omega)$ for a given state ω, can be 1 or 0 or "undefined".

In a short paper P. Busch [143, p1] observed, "it is possible to formulate conditions (such as restrictions of the set of states, or superselection rules) under which a quantum system will appear to display (approximately) classical behavior. But a theoretical explanation of why, and under what circumstances, such classical conditions come to be satisfied seems to be lacking." Boolean fractions can illuminate this issue since for any fixed condition the system is Boolean, but fractions with different conditions always form a non-Boolean subsystem.

8.2.10 Simultaneous Physical Measurements and Indeterminacy

To see how conditionals might represent the subtleties of simultaneous measurements, suppose b is a proposition or event describing the experimental conditions and apparatus for measuring the position of a particle q, and let a be the proposition or event representing the act of measuring the position of q. So (a|b) is the conditional referring to the position measurement of q given the experimental conditions for measuring that position. Similarly, let (c|d) be the conditional representing the act c of measuring the velocity of q given the experimental conditions d for measuring the velocity of q. Therefore if we wish to accomplish both a measurement of position and of velocity under the appropriate conditions, we want (a|b) ∧ (c|d). Now suppose there were a state or condition s such that b ∧ s = d ∧ s ≠ 0. That is, suppose (b|s) = (d|s) ≠ (0|s). In other words, suppose that given state s, b=d ≠ 0. Then conditioning (a|b) ∧ (c|d) by s easily yields ((a|b) ∧ (c|d) | s) = ((a|b)|s) ∧ ((c|d)|s) = (a|bs) ∧ (c|ds) = (a|bs) ∧ (c|bs) = (a ∧ c | bs), indicating that a and c can be measured under the joint condition b ∧ s.

However if there is no such state s, that is, if b ∧ d = 0, then in any state s in which condition b is satisfied (s ≤ b) condition d cannot be satisfied because s ∧ d ≤ b ∧ d = 0. So measuring position yields ((a|b) ∧ (c|d) | b) = ((a|b)|b) ∧ ((c|d)| b) = (a|b) ∧ (c|db) = (a|b) ∧ U = (a|b). And measuring velocity similarly yields (c|d). But trying to measure both position and velocity by satisfying both measurement conditions yields ((a|b) ∧ (c|d) | bd) = (a|bd) ∧ (c|bd) = U ∧ U = U. One or the other condition, b or d, can first be imposed on (a|b) ∧ (c|d) to get a measurement, but as soon as one condition is satisfied the other can't be satisfied. This is the source of non-commutativity and non-distributivity of quantum measurements.

8.2.11 Quantum Logic
Varadarajan [144, p105] adopted the structure of an orthocomplemented lattice as the basic algebraic construct for quantum logic. His formulation nicely characterizes the abstract algebraic situation as follows:

8.2.12 Definition of Orthocomplementation
An orthocomplementation of a lattice L under partial order < is a mapping $(\perp)$: $a \to a^\perp$ of L into L such that

i) $\perp$ is one-to-one and maps L onto itself,
ii) $a < b$ implies $b^\perp < a^\perp$,
iii) $a^{\perp\perp} = a$ for all a,
iv) $a \wedge a^\perp = 0$ for all a,
v) $a \vee a^\perp = 1$ for all a.

Varadarajan noted that (ii)-(iv) imply (i) and (v).

8.2.13 Definition of Orthogonality for Conditionals
Two conditionals (a|b) and (c|d) are said to be *orthogonal* (disjoint) if $(a|b) \wedge (c|d) = (0 \mid b \vee d)$. This will be expressed as $(a|b) \perp (c|d)$. The set $(a|b)^\perp = \{(c|d): (c|d) \perp (a|b)\}$ is the set of all conditionals orthogonal to (a|b).

Expressed in terms of the components of the propositions, $(a|b) \perp (c|d)$ is easily equivalent to having $ab \le c'd$ and $cd \le a'b$: ($0 = abd' \vee b'cd \vee abcd = ab(d' \vee cd) \vee (b' \vee ab)cd \Leftrightarrow ab(d' \vee cd) = 0$ and $ab(d' \vee cd) = 0$. Therefore $ab \le (d' \vee cd)' = c'd$, and similarly for $cd \le a'b$.)

In [69] (and earlier too) the deductive relation $\le_{pm}$ was defined, and it was shown that $(a|b) \le_{pm} (c|d)$ just in case $ab \le cd$ and $c'd \le a'b$. So (a|b) and (c|d) are orthogonal in case $(a|b) \le_{pm} (c|d)'$. ["pm" refers to "probabilistically monotonic".] The relation $ab \le cd$ has been denoted $(a|b) \le_{tr} (c|d)$, meaning the truth of (a|b) implies the truth of (c|d). The relation $c'd \le a'b$ is equivalent to $a \vee b' \le c \vee d'$, which has been referred to as the deductive relation (a|b) $\leq_{nf}$ (c|d), meaning the non-falsity of (a|b) implies the non-falsity of (c|d).

8.2.14 Theorem on Orthogonality
The set of conditionals orthogonal to (a|b) is (a|b)$^\perp$ = {(a'bx | ab ∨ y): x, y in $\mathcal{B}$}.

Proof of Theorem 8.2.14 (a|b)$^\perp$ = {(c|d): (c|d) ⊥ (a|b)} = {(c|d): (a|b) ∧ (c|d) = (0|b ∨ d)}. So (abd' ∨ b'cd ∨ abcd = 0. Therefore abd' = 0, b'cd = 0, and abcd = 0. So ab ≤ d, cd ≤ b, cd ≤ (ab)'. Combining the last two inequalities yields cd ≤ b(ab)' = a'b, which implies both of them. So cd = a'bx, for some x in $\mathcal{B}$ and d = ab ∨ y, for some y in $\mathcal{B}$. So (c|d) = (a'bx | ab ∨ y) for some x, y in $\mathcal{B}$. That is, (a|b)$^\perp$ = {(a'bx | ab ∨ y): x, y in $\mathcal{B}$}. That completes the proof of the theorem.

8.2.15 Theorem on Orthogonal Closure
The set (a|b)$^\perp$ = {(c|d): (c|d) ⊥ (a|b)} of all conditionals orthogonal to (a|b) is closed under disjunction and conjunction.

Proof of Theorem 8.2.15 If (c|d) ∈ (a|b)$^\perp$ and (e|f) ∈ (a|b)$^\perp$ then (c|d) ∨ (e|f) = (cd ∨ ef | d ∨ f) ∈ (a|b)$^\perp$ just in case cd ∨ ef ≤ a'b and ab ≤ (cd ∨ ef)'(d ∨ f). Since (c|d) ∈ (a|b)$^\perp$ therefore cd ≤ a'b and ab ≤ c'd. Similarly ef ≤ a'b and ab ≤ e'f. So cd ∨ ef ≤ a'b and ab ≤ (c'd)(e'f) ≤ (cd ∨ ef)'(d ∨ f). Therefore [(c|d) ∨ (e|f)] ⊥ (a|b). The conjunction (c|d) ∧ (e|f) = (c ∨ d')(e ∨ f') | (d ∨ f) = (cdf' ∨ d'ef ∨ cdef | d ∨ f) ∈ (a|b)$^\perp$ just in case cdf' ∨ d'ef ∨ cdef ≤ a'b and ab ≤ [(c ∨ d')(e ∨ f')]' (d ∨ f). Since cd ≤ a'b and ef ≤ a'b therefore cdf' ∨ d'ef ∨ cdef ≤ a'b. Now [(c ∨ d')(e ∨ f')]'(d ∨ f) = [(c ∨ d')' ∨ (e ∨ f')'](d ∨ f) = c'd ∨ e'f. Since ab ≤ c'd, therefore ab ≤ (c'd) ∨ (e'f). Therefore [(c|d) ∧ (e|f)] ⊥ (a|b). That completes the proof.

If A and B are two Boolean events and AB = 0 = A'B' then B = A'. Similarly, it is easy to show that if (a|b) ∧ (c|d) = (0| b ∨ d) = (a|b)' ∧ (c|d)' then (c|d) = (a|b)'. Note that just as A'B' = 0 means A ∨ B = 1, for conditionals, $(a|b)' \wedge (c|d)' = (0|\ b \vee d)$ means $(a|b) \vee (c|d) = (1|\ b \vee d)$.

Now, although the algebra of Boolean fractions under the disjunction, conjunction and negation operations of Section 8.2.2 does not form a lattice, it almost does, and having a full lattice does not seem to be necessary to capture the full quantum dynamics. Concerning the algebra $(\mathcal{B}|\mathcal{B})$ of Boolean fractions of a Boolean algebra $\mathcal{B}$ there is the following result that parallels Varadarajan:

8.2.16 Complementation in $(\mathcal{B}|\mathcal{B})$

A conditional event algebra $(\mathcal{B}|\mathcal{B})$ has a relative complement $(a|b)'$ for each member $(a|b)$ in $(\mathcal{B}|\mathcal{B})$ such that

1) $(')$ is one-to-one and onto $(\mathcal{B}|\mathcal{B})$,
2) $(a|b) \leq_{pm} (c|d) \Rightarrow (c|d)' \leq_{pm} (a|b)'$,
3) $((a|b)')' = (a|b)$,
4) $(a|b) \wedge (a|b)' = (0|b)$ for all $(a|b)$ in $(\mathcal{B}|\mathcal{B})$,
5) $(a|b) \vee (a|b)' = (1|b)$ for all $(a|b)$ in $(\mathcal{B}|\mathcal{B})$.

Using the operations on conditionals these are all very easy to show.

8.3 Quantum Logic & Conditional Events

From a purely logical point of view, some pairs of quantum events or pairs of quantum propositions cannot be simultaneously verified. This notion of simultaneous verifiability has been well expressed algebraically by V.S. Varadarajan [144, p118]:

8.3.1 Definition of Simultaneous Verifiability

Two members "a" and "c" of a logic L are *simultaneously verifiable*, $a \leftrightarrow c$, if there are members a_1, c_1, and e of L such that $a_1 \wedge c_1 = 0$ and $a_1 \wedge e = 0$ and $e \wedge c_1 = 0$, and with $a = a_1 \vee e$ and $c = e \vee c_1$.

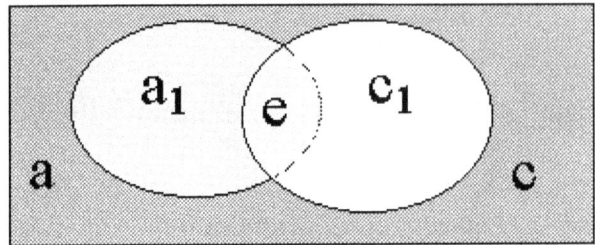

Figure 8.1 Simultaneous Verifiability in an Orthocomplemented Lattice

Obviously, in a Boolean algebra any two members "a" and "c" are always simultaneously verifiable. However in conditional event algebra only some conditionals are simultaneously verifiable:

8.3.2 Theorem on Simultaneous Verifiability
In the conditional event algebra two conditionals (a|b) and (c|d) are *simultaneously verifiable*, (a|b) ↔ (c|d), if and only if $ab \leq d$ & $cd \leq b$.

That is, two conditionals are simultaneously verifiable if and only if the truth of one conditional implies the applicability of the other conditional. That is, two conditionals are simultaneously verifiable if verifying one implies that the other is applicable and so can be verified.

Proof of Theorem 8.3.2 Let (a|b) and (c|d) be two conditionals. If (a|b) ↔ (c|d) then there exist conditionals (α|β), (χ|δ) and (e|f) such that

$$(\alpha|\beta) \wedge (\chi|\delta) = (0 \mid \beta \vee \delta), \quad (8.3.1)$$
$$(\alpha|\beta) \wedge (e|f) = (0 \mid \beta \vee f), \quad (8.3.2)$$
$$(\chi|\delta) \wedge (e|f) = (0 \mid \delta \vee f), \quad (8.3.3)$$

and

$$(a|b) = (\alpha|\beta) \vee (e|f) \quad (8.3.4)$$
$$(c|d) = (\chi|\delta) \vee (e|f) \quad (8.3.5)$$

Using the conditional operations and the definition of equivalence of conditionals, equation (8.3.1) is equivalent to

$$\alpha\beta\delta' \vee \beta'\chi\delta \vee \alpha\beta\chi\delta = 0,$$

which simplifies by Boolean operations and a de Morgan formula to two inequalities:

$$\alpha\beta(\delta' \vee \chi\delta) \vee (\beta' \vee \alpha\beta)\chi\delta = 0,$$
$$\alpha\beta(\delta' \vee \chi\delta) = 0 = (\beta' \vee \alpha\beta)\chi\delta,$$
$$\alpha\beta \leq \chi'\delta \text{ and } \chi\delta \leq \alpha'\beta. \tag{8.3.6}$$

Similarly, equations (8.3.2) and (8.3.3) are equivalent to

$$\alpha\beta \leq e'f \text{ and } ef \leq \alpha'\beta, \tag{8.3.7}$$
$$\chi\delta \leq e'f \text{ and } ef \leq \chi'\delta. \tag{8.3.8}$$

Equations (8.3.4) and (8.3.5) are respectively equivalent to the following two lines:

$$ab = \alpha\beta \vee ef \text{ and } b = \beta \vee f, \tag{8.3.9}$$
$$cd = \chi\delta \vee ef \text{ and } d = \delta \vee f \tag{8.3.10}$$

Starting with (8.3.9),

$$\begin{aligned} b = \beta \vee f &= \alpha\beta \vee \alpha'\beta \vee ef \vee e'f \\ &= ab \vee \alpha'\beta \vee e'f \\ &\geq ab \vee \chi\delta \\ &\geq \chi\delta \end{aligned}$$

using first (8.3.9) and half of both (8.3.6) and (8.3.8).

Since also $b = \beta \vee f \geq f \geq ef$, therefore $b \geq \chi\delta \vee ef = cd$. That is, $cd \leq b$. By symmetry, $ab \leq d$. Therefore both $ab \leq d$ and $cd \leq b$. That proves the first half of Theorem 8.3.2.

Conversely, suppose (a|b) and (c|d) are two conditionals such that $ab \leq d$ and $cd \leq b$. Let

$$(\alpha|\beta) = (ac'|b),$$
$$(\chi|\delta) = (a'c|d),$$
$$(e|f) = (ac|bd).$$

Then

$$(\alpha|\beta) \wedge (e|f) = (ac'|b) \wedge (ac|bd)$$
$$= ((ac')(b)(bd)' \vee (b')(ac)(bd) \vee (ac')(b)(ac)(bd) \mid b \vee bd)$$
$$= (ac'bd' \vee 0 \vee 0 \mid b) = (0|b)$$

using that ab ≤ d, that is, abd' = 0. So (α|β) and (e|f) are orthogonal. By symmetry, so too are (χ|δ) and (e|f). Also (α|β) and (χ|δ) are orthogonal because (α|β) ∧ (χ|δ) = (ac'|b) ∧ (a'c|d) = (ac'bd' ∨ b'a'cd ∨ ac'ba'cd | b ∨ d) = (0 | b ∨ d), using that both abd' = 0 and b'cd = 0.

Finally, (α|β) ∨ (e|f) = (ac'|b) ∨ (ac|bd) = (ac'b ∨ acbd | b ∨ bd) = (ac'b ∨ abc | b) = (a|b), using that ab ≤ d in the second to last step. So (a|b) = (α|β) ∨ (e|f).

By symmetry, (c|d) = (χ|δ) ∨ (e|f). Therefore (a|b) ↔ (c|d). That completes the proof of Theorem 8.3.2.

8.3.3 Corollary on Simultaneous Verifiability

$$(a|b) \leftrightarrow (c|d) \text{ iff } (a|b) \wedge (c|d) = (abcd | b \vee d).$$

The corollary follows easily by computing (a|b) ∧ (c|d) using the conjunction operation and noting that the terms abd' and cdb' are both 0, leaving the term abcd alone in the consequent.

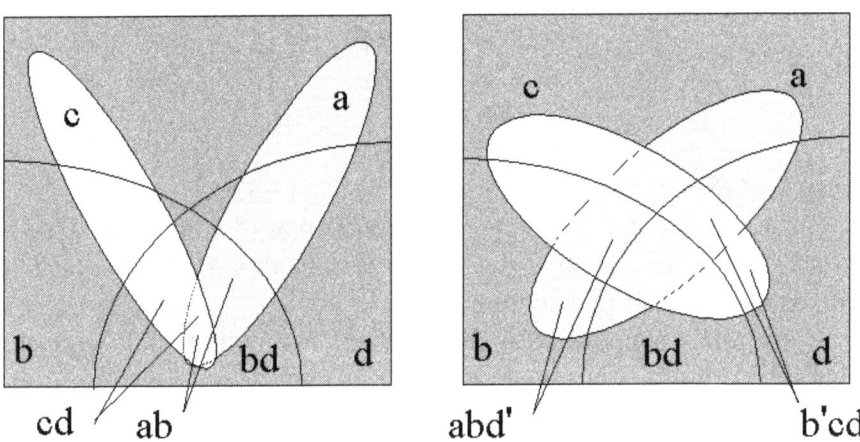

Figure 8.2 Left: Simultaneous Verifiability for Conditional Events (a|b) & (c|d)

If (a|b) and (c|d) are *simultaneously verifiable* then according to the theorem the truth of either conditional implies the applicability of the other conditional, and so if either is true then the other can be verified as either true or false. However if one conditional is false, nothing can be said about the other conditional. The following definition remedies this situation.

8.3.4 Definition (Simultaneous Falsifiability)
Two conditionals (a|b) & (c|d) are *simultaneously falsifiable* if and only if their negations, (a'|b) & (c'|d), are simultaneously verifiable. That is, (a'|b) ↔ (c'|d).

8.3.5 Corollary on Simultaneous Falsifiability
In the conditional event logic two conditionals (a|b) & (c|d) are simultaneously falsifiable if and only if $a'b \le d$ and $c'd \le b$, that is, if the falsity of one conditional implies the applicability of the other conditional.

8.3.6 Corollary on Simultaneous Verifiability and Falsifiability
Two conditionals (a|b) & (c|d) are simultaneously verifiable and simultaneously falsifiable if and only if b = d.

Proof of Corollary 8.3.6 By the Theorem $ab \le d$ and $cd \le b$, and also $a'b \le d$ and $c'd \le b$. So obviously $b = ab \vee a'b \le d$ and similarly $d = cd \vee c'd \le b$. So b=d. That completes the proof of the corollary.

Two conditionals, (a|b) & (c|d), will be called *compatible* (≈) in case both (a|b) ↔ (c|d) and (a|b)' ↔ (c|d)', that is in case they are simultaneously verifiable and simultaneously falsifiable.

In various ways many authors [145; 146, p133; 148, p25; 107, p.293; 92, p120], have expressed the fact that quantum propositions are completely classical if and only if they reside in a common Boolean sub-algebra of the larger non-Boolean system of all quantum propositions. As the next two results will show, for conditional event algebra this corresponds to conditional propositions with equivalent conditions.

8.3.7 Theorem on Boolean Sub-Algebras
Two conditionals (a|b) and (c|d) are in a common Boolean sub-algebra of the (non-Boolean) algebra of conditionals if and only if $b = d \ne 0$.

Proof of Theorem 8.3.7 Suppose (a|b) & (c|d) are in a common Boolean algebra **A**. Then neither b nor d can be 0 since otherwise U = (1|0) would be in **A**. But if U ∈ **A**, then U has a complement (negation) X such that X ∨ U = 1 and X ∧ U = 0. But from the operations on conditionals, or from the 3-valued truth tables of Table 1, X ∨ U = X and X ∧ U = X. Therefore 1 = X ∨ U = X = X ∧ U = 0, which by definition is impossible in a non-degenerate Boolean algebra because 1 ≠ 0.

Since b ≠ 0, then suppose ω is an instance of b taken as a Boolean event (or as an atom in the logical terminology). That is, suppose ω ∈ b. Now either ω ∈ a or ω ∈ a′. If ω ∈ a then consider the expression

$$(a|b) \wedge [(a'|b) \vee (c|d)]$$

but if ω ∈ a′ then instead consider the expression (a′|b) ∧ [(a|b) ∨ (c|d)].

For ω, the appropriate expression has truth-value T ∧ [F ∨ t(c|d)], where t(c|d) denotes the truth-value of (c|d) for ω. By cases it follows easily that

$$T \wedge [F \vee t(c|d)] = \begin{cases} F, \text{ if } t(c\,|\,d) = F \text{ or } U, \\ T, \text{ if } t(c\,|\,d) = T. \end{cases}$$

On the other hand, [(a|b) ∧ (a′|b)] ∨ [(a|b) ∧ (c|d)], or instead [(a′|b) ∧ (a|b)] ∨ [(a′|b) ∧ (c|d)] in case ω ∈ a′, has truth-value

$$[T \wedge F] \vee [T \wedge t(c|d)] = \begin{cases} F, \text{ if } t(c\,|\,d) = F, \\ T, \text{ if } t(c\,|\,d) = T \text{ or } U. \end{cases}$$

But these two expressions must be equal because **A** is a (distributive) Boolean algebra. So t(c|d) ≠ U. Thus ω ∈ d. Thus any ω in b is also in d. That is, b ≤ d. By symmetry, d ≤ b. Therefore b=d.

Conversely, if b=d, then (a|b) and (c|d) are in the common Boolean algebra (*B*|b) = {(x|b): x ∈ *B*} since the operations on conditionals all reduce to Boolean operations when the antecedents are equal, and

(1|b) and (0|b) are the unity and zero elements of $(\mathcal{B}|b)$. That completes the proof of the Theorem 8.3.7.

Note that if b=d=0, then both (a|b) & (c|d) are equivalent to (1|0), the totally inapplicable conditional U. But U cannot be in a non-degenerate (1≠0) Boolean algebra because $U' = U$. In a Boolean algebra, $c \vee c' = 1$ and $c \wedge c' = 0$ for any member c. Thus, if $U' = U$ then $1 = U \vee U' = U \vee U = U \wedge U = U \wedge U' = 0$, which means that U would be the only member. That is, all members are equivalent to U. So again (a|b) & (c|d) would be in the common (non-Boolean, degenerate) algebra {U} with just one member.

8.3.8 Corollary on Boolean Sub-Algebras
Two conditionals (a|b) & (c|d) are simultaneously verifiable and falsifiable if and only if they are in a common Boolean sub-algebra.

If (a|b) and (c|d) are two conditionals and $b \leq d$, then (c|d) is verifiably true or false whenever (a|b) is true or false. That is, if (a|b) is applicable then (c|d) is applicable. This has been denoted (a|b) $\leq_{ap}$ (c|d) in [69] as part of a hierarchy of eleven deductive relations on conditionals.

The deductive relation $\leq_{bo}$ was defined here and in [66] & [69] by (a|b) $\leq_{bo}$ (c|d) if and only if b=d and $ab \leq cd$. As such, $\leq_{bo}$ is just Boolean deduction restricted to conditionals with equivalent antecedents. This appears to be the right deductive relation for quantum logic. It is Boolean deduction in the Boolean algebra $(\mathcal{B}|b) = \{(a|b): a \in \mathcal{B}\}$.

8.3.9 Theorem on Uniqueness of Relative Negation
Let (a|b) and (c|d) be two conditionals. Then (a|b) $\wedge$ (c|d) = (0 | b $\vee$ d) and (a|b) $\vee$ (c|d) = (1 | b $\vee$ d) if and only if b=d and (c|d) = (a|b)'.

Proof of theorem 8.3.9 Using the operations, the first equation is equivalent to $abd' \vee b'cd \vee abcd = 0$, and the second equation is equivalent to $ab \vee cd = b \vee d$. So $abd' = 0$, $b'cd = 0$, and $abcd = 0$. Conjunction on both sides of the equation ($ab \vee cd = b \vee d$) by d' yields $0 = bd'$. So $b \leq d$. By symmetry $d \leq b$. Therefore b=d. So $ab \vee cb = b$. But ($ab \vee cb = b$) $\Leftrightarrow$ ($b \leq a \vee c$) $\Leftrightarrow$ ($ba'c' = 0$) $\Leftrightarrow$ $a'b \leq c$. Also ($abc = 0$) $\Leftrightarrow$ ($c \leq (ab)'$) $\Leftrightarrow$ ($c \leq a'b \vee b'$). Therefore, $a'b \leq c \leq a'b$

v b'. Since c is between a'b and (a'b v b'), c = a'b v xb' for some proposition x. So (c|d) = (a'b v xb' | b) = (a'b v xb'b |b) = (a'|b) = (a|b)'. Conversely, if b=d and (c|d) = (a|b)', then clearly (a|b) ∧ (c|d) = (a|b) ∧ (a|b)' = (0|b) = (0 | b v d) and (a|b) v (c|d) = (a|b) v (a|b)' = (1|b) = (1 | b v d). That completes the proof of Theorem 8.3.9.

8.3.10 Hilbert Space and Other Formulations
Indicating a still rich field of research Coecke et al [147, p17] listed two paragraphs of recent efforts by various authors to generalize orthoalgebras and lattices.

They defined (p16) an orthoalgebra as a pair $(L, \oplus)$ where L is a set and $\oplus$ is a commutative, associative, partial operation on L such that

(1) There exists $0 \in L$ such that $\forall p \in L, p \oplus 0 = p$,
(2) There exists $1 \in L$ such that $\forall p \in L$, there exists a unique $q \in L$ such that $p \oplus q = 1$,
(3) $p \oplus p = 0$ if it exists.

In the algebra of conditionals there is a zero for each condition, which makes the above framework come out a little differently.

8.3.11 Theorem on Orthoalgebras
If (a|b) ⊕ (c|d) is defined to be the conditional event (abc'd v a'bcd | b v d) then

(1) (a|b) ⊕ (0|b) = (a|b),
(2) (a|b) ⊕ (a'|b) = (1|b), and (a'|b) is unique, and
(3) (a|b) ⊕ (a|b) = (0|b).

Proof of Theorem 8.3.11 Except for uniqueness in (2), items (1)–(3) of the Theorem are obvious from the definition of ⊕. Suppose that (c|d) is another complement of (a|b) satisfying (2). That is, suppose (a|b) ⊕ (c|d) = (1|b). Then (abc'd v a'bcd | b v d) = (1|b). So by the definition of equal conditionals, b v d = b and abc'd v a'bcd = b. Since b v d = b, therefore d ≤ b. Now abc'd v a'bcd = b is equivalent to (ac' v ca')bd = b, which means b ≤ (ac' v ca')d. This is equivalent to b[(ac' v ca')d]' = 0. That is, 0 = b[(ac' v ca')' v d'] = b[(a' v c)(c' v a) v d'] = b[a'c' v ac v d'] = a'bc' v abc v bd'. Therefore a'bc' = 0,

$abc = 0$, and $bd' = 0$. But $bd' = 0$ means $b \le d$. So $b=d$. Furthermore, $(a'bc' = 0)$ is equivalent to $(a'b \le c)$, and $(abc = 0)$ is equivalent to $c \le (ab)' = (a' \vee b') = (a'b \vee b')$. Combining these results about c, we have $a'b \le c \le (a'b \vee b')$. But then $(c|d) = (a'b \vee xb' \mid b)$ for some $x \in \mathcal{B}$. So $(c|d) = (a'b \vee xb' \mid b) = (a'b \vee xb'b \mid b) = (a'b|b) = (a'|b)$. Therefore uniqueness is proved, and that completes the proof of the theorem.

8.3.12 Orthocomplementations

Piron [148, p8] following Varadarajan [144] defined a lattice as orthocomplemented if there is a mapping that assigns to each element "a" of the lattice a negation element a' such that for any elements a and c:

1) $(a')' = a$,
2) $a \wedge a' = 0$ and $a \vee a' = 1$,
3) $a \le c$ implies $c' \le a'$.

This definition is expanded (p30) to a "relative" orthocomplementation by restricting all elements to some fixed element b so that

4) $(a')'b = ab$,
5) $ab \wedge a'b = 0$ and $ab \vee a'b = b$,
6) $ab \le cb$ implies $c'b \le a'b$.

These relationships are all special cases of the relative complementation in $(\mathcal{B}|\mathcal{B})$ as expressed in Section 8.2.20. That is, the double negation of equation 4) above is implied by equation 3) of Section 8.2.20. Statement 5) above is equivalent to the combination of 4) and 5) of Section 8.2.20. And statement 6) above is implied by 2) of Section 8.2.20 by letting $b = d$.

8.3.13 Distinctions Possible with Boolean Fractions

It is significant that conditionals (Boolean fractions) can often distinguish relationships that are not distinguishable in the orthodox quantum formulations. For instance, it is easy to see that $(b_1|b_2)' \approx (a_1|a_2)$ is equivalent to $\{b_1'b_2 \le a_2$ and $a_1a_2 \le b_2\}$ which means that the truth of $(a_1|a_2)$ implies the applicability of $(b_1|b_2)$ and the falsity of $(b_1|b_2)$ implies the applicability of $(a_1|a_2)$. Thus one might wish to discuss several kinds of simultaneous verifiability for two conditionals $(a_1|a_2)$ and $(b_1|b_2)$ such as:

1) $a_1a_2 \le b_2$; if $(a_1|a_2)$ is true then $(b_1|b_2)$ is applicable (and so verifiable).
2) $a_1'a_2 \le b_2$; if $(a_1|a_2)$ is false then $(b_1|b_2)$ is applicable (and so verifiable).
3) $a_1a_2 \le b_2$ and $b_1b_2 \le a_2$; if either conditional is true then the other is applicable. In this book this has been called *simultaneous verifiability* and denoted by $(a_1|a_2) \leftrightarrow (b_1|b_2)$.
4) $b_1'b_2 \le a_2$ and $a_1'a_2 \le b_2$; if either conditional is false then the other is applicable. In this book this has been called *simultaneous falsifiability*: $(a_1|a_2)' \leftrightarrow (b_1|b_2)'$.
5) $a_1'a_2 \le b_2$ and $b_1b_2 \le a_2$; if the first conditional is false then the second conditional is applicable, and if the second conditional is true then the first is applicable. This combines the relationships in 1) and 2), reversing the notation in 2). This is $(a_1|a_2)' \leftrightarrow (b_1|b_2)$.
6) $a_2 \le b_2$; if $(a_1|a_2)$ is applicable then $(b_1|b_2)$ is applicable (and so verifiable). This is 1) and 2).
7) $a_2 = b_2$; equivalent conditions. This has been called *compatibility*. Note $(a_1|a_2)' \leftrightarrow (b_1|b_2)$ and $(a_1|a_2) \leftrightarrow (b_1|b_2)'$ both hold if and only if b=d. Also 3) and 4) hold just in case $a_2 = b_2$.

8.3.14 True, Wholly True and Given True

One of the main problems confronting anyone trying to reformulate quantum mechanics is how the truth of a quantum proposition is defined as compared to Boolean logic or probability logic. While a Boolean proposition b is supposed to be either true or false, with nothing in between, that only applies to each instance, occurrence or atomic state ω. Usually, in general probability theory, a moving occurrence ω determines whether a fixed proposition "occurs" or is "true" depending on whether $\omega \in b$ or $\omega \in b'$. The propositions or events b are temporarily fixed while the instances ω are changing, being any one of a sample space of outcomes. Thus a proposition may be true in one instance and false in another. In the indicator representation of conditionals the truth of b for ω can be written as $b(\omega) = 1$.

Now if b is true for all occurrences ω in the sample space Ω then this can be expressed as "b≡1", or simply b=1. This is the situation when b is a certainty, when b is wholly true.

Still another type of truth for b is "given b", which is a restriction of the set Ω to those instances for which b is true. This can be expressed as (x | b) or (• | b), "given b", meaning any member of the Boolean algebra ($\mathcal{B}$|b) of events x each conjoined with b.

While the probability P(b) of b can be one number, the P(b=1) that b is wholly true can be another number [63, p221][67, p1683], and of course the conditional probability P(• | b) when b is "given" is something else again.

In quantum logic, a pure (atomic) state ω may be considered fixed, so that the truth of a moving proposition b depends upon whether b includes ω. The events can move while the state ω is fixed. Thus, given a pure state ω a quantum proposition b is either wholly true or wholly false depending on whether it happens to include the fixed state ω.

Actually, in practice both the propositions and the states move, especially in quantum mechanics, where determining the truth of a proposition b generally changes the state as a result of the measurement.

8.3.15 Quantum Operations and Truth Revisited

As mentioned earlier, G. Schay [60] observed that the disjunction and conjunction operations on conditionals promoted here seem to correspond to the usual meaning of "or" and "and", but he mostly passes them by for operations that form a lattice. Using the conjunction operation, for instance, to define a partial order in the usual way (that is, b ≤ c means b ∧ c = b) Schay undoubtedly discovered that (a|b) ∧ (c|d) is the GLB of its components but (a|b) ∨ (c|d) is not the LUB. So he chose a different disjunction operation to go with ∧. Similarly, he chose a different conjunction operation to go with ∨.

However, as Schay himself showed, the resulting lattices are distributive, making them inadequate for representing quantum operations. While the properties of a lattice are not essential for representing quantum operations, the property of non-distributivity is.

In their account of the quantum operations Aerts et al [152] carefully explained how a quantum state p represented by a unit vector v_p determines the truth of a quantum proposition b as it is contained (or not) in the closed linear subspace M_b representing b in the Hilbert space. They pointed out that although conjunction and implication behave as they do in the classical case, disjunction of two propositions does not. While a ∧ b corresponds to $M_a \cap M_b$ and a ≤ b (a implies b) corresponds to $M_a \subset M_b$, the disjunction a ∨ b corresponds to cl($M_a \cup M_b$), the *closure* of the union of the closed subspaces M_a and M_b. This closed subspace is the LUB of M_a and M_b. It is likely that Schay [60] was influenced by this LUB connection when he chose his operations on conditionals. However it is possible to recreate similar relationships in the algebra of conditionals while maintaining the intuitive operations promoted in this book.

The closed quantum subspaces M_b for propositions b always includes the zero vector 0 as well as the negations of all members. They are subspaces not ideals or deductively closed sets. These subspaces M_b seems to correspond to the conditional subspaces (𝐵|b) for any given fixed proposition b. For two conditional spaces, (𝐵|b) and (𝐵|d), their LUB would be the subsystem {(𝐵|b) ∪ (𝐵|d) ∪ (𝐵 | b ∨ d)}, of conditionals which is closed under conjunction, disjunction and negation (but not conditioning) and in general is no longer a Boolean algebra.

This disjunction of two quantum propositions b and d therefore refers to their disjunction as given conditions: (•|b) ∨ (•|d) = {(•b ∨ •d | b ∨ d)} = {(a|b) ∨ (c|d): any a, c in 𝐵} = {(ab ∨ cd | b ∨ d): any a, c in 𝐵}, something impossible to represent with purely Boolean propositions. The probability of any element of (ab ∨ cd | b ∨ d) is a superposition of the conditional probabilities of (a|b) and (c|d) as given in Section 8.2.12.

> One is well advised to beware of probability statements expressed in the form P(X) instead of P(X | C). The second argument may be safely omitted only if the conditional event or information is clear from the *context*, and is constant throughout the problem. This is not the case in the double slit example.
> L. E. Ballentine [127], "Probability theory in quantum mechanics", 1986

Chapter 9

The Logic of Quantum Measurements in terms of Conditional Events

> We are therefore faced with a breakdown of our customary ideas about indefinite analyzability of each process into various parts, located in definite regions of space and time.
> D. Bohm [117], *Wholeness and the Implicative Order*, p. 73, 1980

9.0 Introduction

This chapter first briefly reviews the character of the quantum measurements that resulted in the development of the standard non-Boolean Hilbert space formulation of quantum logic. Authors Bell, Bohm and Ballentine express the need for a more explicit representation of conditions in quantum logic to describe incompatible quantum measurements. The 3-valued, non-distributive non-monotonic algebra of Boolean fractions, or conditional events answers this need. This concept of a conditional event (also called a Boolean fraction, conditional proposition, and partially defined indicator function) provides a near-Boolean logical object (a|b) that can also non-trivially carry the conditional probability of "event a given event b". This idea of starting with ordered pairs of events before defining the logical operations between them extends Boolean algebra to the "changing context" algebra of conditional events. The ability to explicitly express and logically manipulate conditional expressions with changing conditions allows quantum measurements, implicitly including the given state of the measuring instruments and environment, to be more faithfully represented. A very simple "die throw" example of two conditionals whose conjunction has greater probability than either component demonstrates the remarkable but intuitively clear way that conditionals can be non-monotonic. Conditional event algebra extends the crucial concept of orthogonality. This allows V.S. Varadarajan's "orthogonal components definition" of the simultaneous verifiability of two quantum propositions to be captured in the simple algebraic fact that the truth of either conditional proposition must imply the applicability of the other conditional proposition. It becomes clear that two simultaneously verifiable conditionals may not be simultaneously falsifiable. Those conditionals with both properties are compatible, and have equal (equivalent) conditions. Conditionals with equal conditions generate a Boolean sub-algebra of the algebra of conditionals, and all Boolean sub-algebras are of this type. The algebra of conditionals explicates exactly how two incompatible physical measurements can give rise to non-commutative operations in quantum mechanics. New results include a definition of completeness for a set of orthogonal conditional events and a theorem characterizing it. Every set of orthogonal conditionals can be completed, and the conditions in any complete orthogonal set can be uniformly restricted and remain complete. Any conditional (e | f) can be represented in several ways as a disjunction of (e | f) with the orthogonal members of a complete set. Some of these extend to a corresponding formula for the conditional probability of (e | f) in terms of summations of conditional probabilities. The chapter ends with an account of certain "consequence" logics that formulate desirable abstract properties that any deduction function should have if it is to adequately identify the combined consequences $C(A)$ of a set A of possibly non-Boolean propositions. These consequence logics are related to quantum logic in Hilbert space and to the theory of deduction in the algebra of uncertain conditionals. Also see Calabrese [73].

9.1 Principles of Quantum Mechanics
The physical mechanics of objects of very small mass or energy includes the following basic principles:

a) There is a smallest unit of mass or energy. Mass and energy levels are therefore discrete, multiples of these smallest values, not continuous variables.

b) Quantum observation generally alters the state of the object observed. Quantum observables are operators, not just passive variables being measured. The act of determining the value of any one variable associated with a quantum object may put the whole quantum system into a different state where there are new probabilities for variables equaling their respective values.

c) Experimental setup conditions for the measurement of one variable may preclude the conditions necessary for the measurement of another variable. Since the value of one variable can be changed by the measurement of another variable, simultaneous measurements are sometimes impossible. For example, a measurement of the position of an electron alters its velocity, and a measurement of its velocity alters its position. The experimental setup to precisely determine them simultaneously is impossible.

d) A particle appears to have an associated wave that extends into space in all directions allowing it to move according to "inference probabilities" such as displayed in the double-slit experiments.

Of these four principles, this chapter will mainly explore the consequences of b) and c), but the results are also consistent with a) and d).

9.1.1 Formulation of the Standard Algebra of Quantum Measurements

As H. Putnam put it [104, p51], "The whole function of the linear spaces used in quantum mechanics is to provide a convenient mathematical representation of the lattice of physical propositions... ...the corresponding lattice of physical propositions is not 'Boolean'; in particular, distributive laws fail." But Boolean logic is, of course, completely distributive.

Looking for a mathematical system that could represent quantum mechanics, J. von Neumann and G. Birkhoff [115] formulated "orthodox" quantum logic – a complete, normed, inner-product space, also an orthocomplemented lattice. It is no secret that people have been looking for a more natural formulation ever since.

But what would have happened had the (non-distributive) algebra of Boolean fractions (conditional events) been available 100 years ago? Perhaps a much more natural expression of quantum logic would have resulted (see Chapter 8 and [72]).

9.1.2 The Need for Explicit Conditions

In discussing the standard quantum formalism, physicist D. Bohm [117, p74] wrote, "As a simple analysis shows, the impossibility of theoretically defining two non-commuting observables by a single wave function is matched exactly, and in full detail, by the impossibility of the operation together of two overall set-ups that would permit the simultaneous experimental determination of these two variables. This suggests that the non-commutativity of two operators is to be interpreted as a mathematical representation of the incompatibility of the arrangements of apparatuses needed to define the corresponding quantities experimentally."

To debunk the supposed violation of the additive law of probabilities in the well-known double slit experiment, Ballentine [127, p887], focused on the importance of explicit conditions. "... beware of probability statements expressed as P(X) instead of P(X|C). The second argument may be safely omitted only if the conditional event or information is clear from the context and *constant* throughout the problem. This is not the case in the double slit example." Distinguishing between the 3 different slit conditions Ballentine showed how to express with just classical conditional probability theory, the strange x-position (quantum interference) probabilities manifested by a particle passing through either of two slits.

J. S. Bell [120, p166] discussing contextuality in quantum logic says that the word "measurements" is misused, that quantum measurements include the combination of system and apparatus, the complete experimental set-up. According to Bell, forgetting this has led people to expect that results of measurements should obey some simple logic in which the apparatus is not mentioned. "The resulting difficulties soon show that any such logic is not ordinary logic. It is my impression that the whole vast subject of 'Quantum Logic' has arisen in this way from the misuse of a word."

Thus the act of measurement including the experimental setup must be an integral part of the logic of quantum mechanics. It cannot be ignored. If one setup precludes a second setup then the order of measurements can change the results. Therefore some non-commutativity must be represented in the logic of quantum measurements. But Boolean algebra is totally commutative.

9.1.3 Compatible Propositions and Boolean Sub-Algebras

Compatible quantum propositions are those that can be determined simultaneously and which together form a Boolean sub-algebra. This has been expressed in different ways by Piron [148, p25], who keys on the distributivity of the generated sub-algebra, and by Jauch [145; 146, p133], who simply states that the generated sub-algebra is Boolean.

Similarly, Suppes [107, p293] says, "If we avoid noncommuting variables in quantum mechanics, then probability is classical."

Just as the inadequacies of Boolean algebra for quantum logic were early recognized, the inadequacies of the so-called "material conditional" in Boolean logic were also apparent even before Kolmogorov [133] formulated probability theory. As has been well documented, he could not identify P(if b then a) with P(either a or not b) as Bertrand Russell had done in the latter's purely 2-valued logical development. For partially true events or propositions, this was a gross distortion. Thus did probability theory give us P(a|b), the probability of a given b, without defining (a|b), "a given b" as a logical element.

9.1.4 Boolean Fractions

Boolean fractions [63] [67], [69] (also called conditional events) are ordered pairs of propositions (a|b) representing "a given b" and sometimes "if b then a". They can be combined with the three familiar Boolean logical operations, "and", "or" and "not", but also allowed are iterations, fractions of fractions, such as "(a given b) given c", "a given (b given c)", and most generally, "(a given b) given (c given d)".

As with our familiar numerical fractions, the joining of two events (or propositions) in an "ordered pair" occurs *before* the operations on events (or propositions) are defined, not afterward. Thus the basic set of objects being used to represent and manipulate information has been greatly expanded. Something similar happens when ordinary numerical fractions are formed as pairs of whole numbers greatly improving the representation and manipulation of distances on a number line.

9.1.5 The Set of Conditionals

We denote by $(\mathcal{B}|\mathcal{B})$ the set of ordered pairs, {(a|b): a, b in a Boolean algebra $\mathcal{B}$}, and call this the set of *conditionals*, "a given b", of $\mathcal{B}$. The proposition or event "b" is called the given *condition* or *context* and the proposition or event "a" is simply the event or proposition specified in that context. "a given b" does not refer to the statement "b implies a" or "b entails a". It can represent "if b then a" providing that this is not necessarily taken to be an implication but instead possibly a partially true statement with a conditional probability. In the domains of logic and probability, (a|b) can represent a logical element "a given b" that can hold the conditional probability P(a|b). With its two components, (a|b) is a new more general and more expressive logical object than a simple Boolean event. (For a full account of deduction with such uncertain conditionals see [69].)

9.2 Three-Valuedness of Conditionals
It was De Finetti [90] who first recognized the 3-valuedness of conditionals.

9.2.1 Definition of Equivalence of Conditionals
Two conditionals (a|b) and (c|d) are equivalent (=), written (a|b) = (c|d), if and only if b=d and ab = cd.

The three logical values of conditionals arise as the components of any conditional (a|b) take on the 2 Boolean truth-values, true (1) or false (0). Because (1|0) = (0|0) by the equivalence relation, just three distinct truth states for conditionals remain: (1|1), (0|1), and (1|0). These correspond respectively to "true" (1), "false" (0) and "inapplicable-undefined" (U). So an arbitrary conditional (a|b) is true, false or inapplicable just in case respectively (ab), (a'b) or b' is true.

The algebra of conditional events (Boolean fractions) provides four operations for "or", "and", "not" and also "given" (conditioning) to handle compound conditional forms such as "((a|b) | (c|d))" for two conditionals (a|b) and (c|d).

9.2.2 Relative Negation
$$\textbf{not} \ (a \ given \ b) = (\textbf{not} \ a) \ given \ b.$$
That is, (a|b)' = (a'| b), and the latter has probability 1 - P(a|b).

9.2.3 Disjunction
"a given b" **or** "c given d" means "ab or cd given either conditional is applicable ". Since (a|b) is true on "ab" and (c|d) is true on "cd", the result of disjunction can be expressed by saying "at least one conditional is true given either is applicable." That is,

$$(a|b) \lor (c|d) = ((ab \lor cd) | (b \lor d)) = (ab \lor cd | b \lor d).$$

9.2.4 Conjunction
"a given b", **and** "c given d" means "at least one conditional is true while the other is not false given either conditional is applicable". That is,

$$(a|b) \land (c|d) = ([ab(c \lor d') \lor (a \lor b')cd] | (b \lor d))$$

$$= (abd' \lor abcd \lor b'cd \mid b \lor d),$$

which also means "in case either conditional is applicable then either they are both true or else one is true while the other is inapplicable."

The algebra $(\mathcal{B}|\mathcal{B})$ of conditionals includes the original Boolean algebra $\mathcal{B}$ as those conditionals (a|1) whose condition is certain. Analogously, these are like the integer fractions whose denominators are 1. Fixing the condition b also yields a Boolean algebra $(\mathcal{B}|b) = \{(a|b): \text{all } a \in \mathcal{B}\}$. It turns out that these are the only Boolean sub-algebras of $(\mathcal{B}|\mathcal{B})$.

9.2.5 Iterated Conditioning

A conditional (c|d) may itself be a condition for another proposition or conditional proposition. Assuming that a subsequent condition is conjoined to the previous ones, and applying the operations we can deduce that ((a|b) **given** (c|d)) = ((a|b) | (c|d)) = (a | b ∧ (c|d)) = (a | (b|1) ∧ (c|d)) = (a | b ∧ (c ∨ d')). So,

$$((a|b) \mid (c|d)) = (a \mid b \land (c \lor d')) = (a \mid b(c \lor d')).$$

The order of preference of the operations is negation (′) before conjunction (∧), before disjunction (∨), before conditioning (|). Conjunction is often expressed by juxtaposition.

The properties of this conditional event algebra can be found in [63] and are completely characterized in [69]. $(\mathcal{B}|\mathcal{B})$ is not distributive except in special cases. (See Section 8.2.4-5 and [72] for necessary and sufficient conditions for both distributive laws.)

9.2.6 Non-Monotonicity of Compound Conditionals

Among the more interesting aspects of how conditionals combine are the non-monotonicities in logic and probability first mentioned by E. Adams [94] and G. Schay [60]. These sound rather unintuitive when they are quoted in the abstract, but simple examples make them very plausible if no less surprising. For instance, in Boolean algebra, P(A and B) ≤ P(A) for any two events A, B. But if A = (a|b) and B = (c|d) then P(A and B) can be greater than both P(A) and P(B). At first this seems strange but consider the following bet in the game of rolling a single normal die once. "Given the roll is less than 5 it will be less than 4, and given the roll is greater than 3 it will be greater than 4".

The component conditionals have conditional probabilities 3/4 and 2/3 respectively, but the conjunction has conditional probability 5/6. This can be calculated using the operations on conditionals, but it is easier to simply check the six possible outcomes of the die roll. They each apply to one or the other of the conditionals, but not to both except for "4", which yields a lone "false" for both. Since the lone false instance "4" contributes "false" to both component conditionals, combining them into one conditional reduces the probability of "false" because that "false" is no longer counted twice, while the context is expanded.

9.3 Simultaneous Verifiability of Conditionals

It is clear that the notion of being orthogonal (disjoint) is an important feature of quantum logic. For conditionals this takes the following form:

9.3.1 Definition of Orthogonality for Conditionals

Two conditionals (a|b) and (c|d) are *orthogonal* (disjoint) if (a|b) ∧ (c|d) = (0 | b ∨ d). This is expressed as (a|b) ⊥ (c|d). Expressed in terms of the components of (a|b) and (c|d) this is equivalent to having ab ≤ c'd and cd ≤ a'b. [This follows from (a|b) ∧ (c|d) = (ab(c ∨ d') ∨ cd(a ∨ b') | b ∨ d) = (0 | b ∨ d) if and only if (ab(c ∨ d') = 0 and cd(a ∨ b') = 0.] So for orthogonal conditionals whenever one conditional is true the other conditional must be false.

Notice that if just one of two orthogonal conditionals (a|b) and (c|d) is applicable for an instance ω, then that conditional will be false for ω because whenever one conditional is true the other conditional must be false, not (merely) inapplicable. But both orthogonal conditionals can be false at ω. And of course they may both be inapplicable at ω.

9.3.2 Definition of Simultaneous Verifiability

In an orthocomplemented lattice V.S. Varadarajan [144, p118] algebraically expressed the fundamental notion of two *simultaneously verifiable* propositions as those whose disjunction could be partitioned into three disjoint and exhaustive propositions. That is, two members "a" and "c" of a logic L are *simultaneously verifiable*, a ↔ c, if there are members a_1, c_1, and e of L such that $a_1 ∧ c_1 = 0$ and $a_1 ∧ e = 0$ and $e ∧ c_1 = 0$, and with $a = a_1 ∨ e$ and $c = e ∨ c_1$.

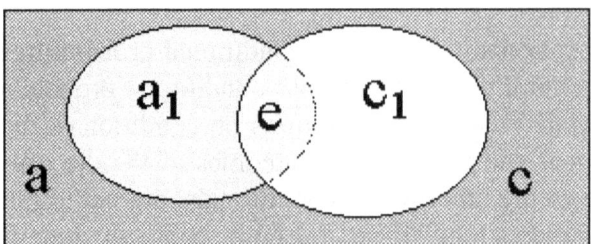

**Figure 9.1 Simultaneous Verifiability
in an Orthocomplemented Lattice**

In a Boolean algebra any two members "a" and "c" are simultaneously verifiable. But for the set of all conditionals there is the following theorem:

9.3.3 Theorem on Simultaneous Verifiability of Conditionals
In the conditional event algebra two conditionals (a|b) and (c|d) are *simultaneously verifiable*, (a|b) ↔ (c|d), if and only if ab ≤ d and cd ≤ b. That is, two conditionals are simultaneously verifiable if and only if the truth of one conditional implies the applicability of the other conditional. That is, two conditionals are simultaneously verifiable if verifying one implies that the other is applicable and so can be verified.

A proof is given in Sections 8.3.2-6 and [72]. It then easily follows from the definition of conjunction of conditionals that

$$(a|b) \leftrightarrow (c|d) \quad \Leftrightarrow \quad (a|b) \wedge (c|d) = (abcd \mid b \vee d).$$

That is, two conditionals are simultaneously verifiable if and only if their conjunction reduces to (abcd | b ∨ d), which is also equivalent to (ac| bd)(bd | b ∨ d).

As called for in the definition of simultaneously verifiable conditionals if (a|b) and (c|d) are simultaneously verifiable, and so ab ≤ d and cd ≤ b, then (a|b) ∨ (c|d) can be partitioned into the disjunction of three orthogonal conditionals:

9.3.4 Orthogonal Expansion Theorem
If (a|b) and (c|d) are simultaneously verifiable, then {(ac'|b), (ac|bd), (a'c|d)} are three orthogonal conditionals for which

$$(a|b) \vee (c|d) = (ac'|b) \vee (ac|bd) \vee (a'c|d)$$

where $(a|b) = (ac'|b) \vee (ac|bd)$ and $(c|d) = (ac|bd) \vee (a'c|d)$, and

$$P((a|b) \vee (c|d)) = P(ac'|b)P(b \mid b \vee d) + P(ac \mid bd)P(bd \mid b \vee d)$$
$$+ P(a'c \mid d)P(d \mid b \vee d).$$

Proof of Theorem 9.3.4 The three conditionals $(ac'|b)$, $(ac|bd)$ and $(a'c|d)$ are easily shown pairwise orthogonal by computing the conjunction of each pair and getting a zero conjunction, using that $abd' = 0$ and $cdb' = 0$. For example,

$$(ac'|b)(ac|bd) = ((ac'b)(bd)' \vee (b')(acbd) \vee (ac'b)(acbd) \mid b \vee bd)$$
$$= ((ac'b)(b' \vee d') \vee (0) \vee (0) \mid b)$$
$$= (abc'd' \mid b) = (0|b).$$

It is also easy to compute the relevant disjunctions to verify the statements of the theorem. For example, $(ac'|b) \vee (ac|bd) = (ac'b \vee acbd \mid b \vee bd) = (ac'b \vee acb \mid b) = (ab|b) = (a|b)$, using that $ab \le d$. By symmetry, or directly, $(c|d) = (ac|bd) \vee (a'c|d)$. So obviously, $(a|b) \vee (c|d) = ((ac'|b) \vee (ac|bd)) \vee ((ac|bd) \vee (a'c|d)) = (ac'|b) \vee (ac|bd) \vee (a'c|d)$. Therefore,

$$P((a|b) \vee (c|d)) = P((ac'|b) \vee (ac|bd) \vee (a'c|d))$$
$$= P(ac'b \vee acbd \vee a'cd \mid b \vee d)$$
$$= P(ac'b \vee acbd \vee a'cd) / P(b \vee d)$$
$$= [P(ac'b) + P(acbd) + P(a'cd)] / P(b \vee d)$$
$$= P(ac'|b)P(b) + P(ac|bd)P(bd) + P(a'c|d)P(d)] / P(b \vee d)$$
$$= P(ac'|b)P(b|b \vee d) + P(ac|bd)P(bd|b \vee d)$$
$$+ P(a'c|d)P(d|b \vee d).$$

That completes the proof of the theorem.

So the probability of the disjunction of two simultaneously verifiable conditionals is the weighted average of the probabilities of the three orthogonal conditionals that combine by disjunction to make it up, where the weights are the relative probabilities of the conditions of the three orthogonal components relative to their disjunction.

This equation parallels similar "superposition" formulas in quantum logic when the state of the system may be either of two overlapping states b or d.

Now if either of two simultaneously verifiable conditionals is true then the truth or falsity of the other one can be determined because it must be applicable, and therefore true or false. But if it is only known that one conditional is false, we would not know whether the other conditional is false versus inapplicable. This motivates the following definition.

9.3.5 Definition of Simultaneous Falsifiability
Two conditionals (a|b) & (c|d) are *simultaneously falsifiable* if and only if their negations, (a'|b) & (c'|d), are simultaneously verifiable. That is, (a'|b) ↔ (c'|d).

Applying this definition to the theorem it easily follows that two conditionals (a|b) & (c|d) are simultaneously falsifiable if and only if $a'b \le d$ and $c'd \le b$, that is, if the falsity of one conditional implies the applicability of the other conditional. Combining the two definitions easily yields the following result:

9.3.6 Simultaneous Verifiability and Falsifiability
Two conditionals (a|b) & (c|d) are simultaneously verifiable and simultaneously falsifiable if and only if b = d.

It turns out that this "equality of conditions" requirement is also necessary and sufficient that both (a|b) and (c|d) are in a common Boolean algebra.

9.3.7 Simultaneous Physical Measurements
Using conditionals we can see how incompatible experimental conditions can lead to non-commutative operations in quantum logic. Let (a|b) be a conditional referring to the position measurement of a particle given the experimental conditions for measuring that position. Similarly, let (c|d) be the conditional representing the act c of measuring the velocity of the particle given the experimental conditions d for measuring that velocity. Wishing to simultaneously measure position and velocity under the appropriate conditions, we need (a|b) ∧ (c|d). Now if bd ≠ 0, then both conditions can be met. So conditioning (a|b) ∧ (c|d) by bd yields ((a|b) ∧ (c|d) | bd) = ((a|b) | bd) ∧ ((c|d) | bd) = (a|bd) ∧ (c|bd) = (ac|bd). So both a and c can be measured simultaneously.

However if bd = 0, there is no common condition that makes most conditionals applicable. If position is first measured by conditioning with b, the result is $((a|b) \wedge (c|d) | b) = ((a|b)|b) \wedge ((c|d)| b) = (a|b) \wedge (c|db) = (a|b) \wedge U = (a|b)$. Similarly, measuring velocity yields (c|d). But try to satisfy both measurement conditions and one gets $((a|b) \wedge (c|d) | bd) = (a|bd) \wedge (c|bd) = U \wedge U = U$. Either of the two conditions, b or d, can be imposed on $(a|b) \wedge (c|d)$ to get a measurement, but if one condition is satisfied the other can't be. This demonstrates the source of non-commutativity and non-distributivity in quantum measurements. Clearly this phenomenon can be well represented in the algebra of conditional events.

9.3.8 Theorem on Generating Boolean Sub-Algebras
Two conditionals (a|b) and (c|d) are in a common Boolean sub-algebra of the (generally non-Boolean) algebra of conditionals if and only if b = d ≠ 0.

(The proof is given in Section 8.3.7-8 and [72] together with a fuller account of many of the preceding results.) Note that the theorem asserts that two conditionals with non-equivalent conditions will always generate a non-Boolean sub-algebra, while two conditionals with equivalent conditions always generates a Boolean sub-algebra. So all Boolean sub-algebras of $(\mathcal{B}|\mathcal{B})$ are of the form $(\mathcal{B}|b) = \{(x|b): x$ any proposition in $\mathcal{B}$, b fixed$\}$. The closed sub-algebra generated by two propositions (a|b), (c|d) will include the two sub-algebras plus what can be deduced from them. For a more complete account of deduction for uncertain conditionals see chapter 6 and [69].

9.3.9 Corollary on Boolean Sub-Algebras
From the last two results concerning conditionals with equivalent conditions it follows in one step that two conditionals (a|b) & (c|d) are simultaneously verifiable and falsifiable if and only if they are in a common Boolean sub-algebra.

Having a common condition for two experimental apparatuses was earlier observed to be essential for the simultaneous measurement of two quantum variables. This feature of quantum mechanics appears to be perfectly represented by the algebra of conditionals because everything is Boolean for simultaneously verifiable propositions, but definitely non-Boolean when two or more of the conditionals have unequal conditions.

9.4 Orthogonality and Conditionals
The concept of orthogonality is central to the Standard Formulation of quantum mechanics in Hilbert Space. To express similar relationships, we further develop the concept of orthogonality in the algebra of conditionals.

9.4.1 Orthogonal Sets of Conditionals
As mentioned earlier, two conditionals (a|b) and (c|d) are defined to be orthogonal if $(a|b) \wedge (c|d) = (0 \mid b \vee d)$, and this is equivalent to having both $ab \leq c'd$ and $cd \leq a'b$, that the truth of one conditional implies the falsity of the other conditional. (Note that orthogonal conditionals are always simultaneously verifiable.)

The latter two inequalities were shown in [69] to be equivalent to having $(a|b) \leq_{pm} (c|d)'$, where the relation $(a|b) \leq_{pm} (c|d)$ was initially defined by the "extended material conditional" deductive equation $(c|d) \vee (a|b)' = (1 \mid b \vee d)$. (This deductive relation is also the weakest probabilistically monotonic (pm) deductive relation.) So in the case of orthogonal conditionals, since $(a|b) \leq_{pm} (c|d)'$ it follows that $P(a|b) \leq P((c|d)')$. Since $P((c|d)') = P(c'|d) = 1 - P(c|d)$, therefore $P(a|b) + P(c|d) \leq 1$.

9.4.2 Definition of a Complete Set of Orthogonal Conditionals (COSC)
Two or more conditionals form an orthogonal set of conditionals by being pairwise orthogonal. A set $\{(a_i|b_i)\}$ of pairwise orthogonal conditionals is *complete* if and only if

$$[(c|d) \perp (a_i|b_i) \text{ for all } i] \Rightarrow (c|d) = (0|d).$$

That is, only zero conditionals are orthogonal to all of the $(a_i|b_i)$.

9.4.3 Lemma for Completeness Theorem
If $J = \{(a_i|b_i)\}$ is a set of pairwise orthogonal conditionals then

$$\vee_i a_i b_i \leq \wedge_i b_i \text{ and } \wedge_i b_i - \vee_i a_i b_i = \wedge_i a_i' b_i.$$

Proof of Lemma 9.4.3 Since J is an orthogonal set of conditionals, $a_i b_i \leq a_j' b_j$, for all $i \neq j$. Therefore $a_i b_i \leq b_j$ for all i, j, and so $\vee_i a_i b_i \leq b_j$ for all j, and so $\vee_i a_i b_i \leq \wedge_j b_j$. The difference, $\wedge_j b_j - \vee_i a_i b_i = (\wedge_j b_j)$

$(\vee_i a_i b_i)' = (\wedge_j b_j)(\wedge_i (a_i b_i)') = \wedge_i (\wedge_j b_j (a_i b_i)') = \wedge_i (\wedge_j b_j (a_i' b_i \vee b_i')) = \wedge_i (\wedge_j b_j (a_i' b_i \vee 0)) = (\wedge_j b_j)(\wedge_i a_i' b_i) = \wedge_i a_i' b_i$. That completes the proof of the lemma.

9.4.4 Completeness Characterization Theorem

An orthogonal set of conditionals J is complete if and only if $\wedge_i a_i' b_i = 0$. That is, an orthogonal set J is complete if there is no instance ω for which all the conditionals of J are false.

Proof of Theorem 9.4.4 Suppose J is complete. Then consider the conditional $(\wedge_i a_i' b_i \mid \vee_i b_i)$. By Definition 9.3.1, it is orthogonal to all the conditionals of J because firstly, $(\wedge_i a_i' b_i)(\vee_i b_i) = (\wedge_i a_i' b_i) \leq a_j' b_j$ for all j. Secondly, $a_j b_j \leq (\wedge_i a_i' b_i)'(\vee_i b_i)$ for all i and j. This follows by using the de Morgan formula to expand $(\wedge_i a_i' b_i)'$ into $\vee_i (a_i b_i \vee b_i')$. Therefore $(\wedge_i a_i' b_i \mid \vee_i b_i)$ is orthogonal to every conditional in J, and J is a COSC. Therefore by completeness, $\wedge_i a_i' b_i = 0$. Conversely, if $\wedge_i a_i' b_i = 0$, then any conditional (c|d) that is orthogonal to all the conditionals of J will satisfy $cd \leq a_i' b_i$ for all i, and so $cd \leq \wedge_i a_i' b_i = 0$. Thus (c|d) = (0|d). So J is complete. That completes the proof of theorem 9.4.4.

In a complete orthogonal set of conditionals $\{(a_i \mid b_i)\}$ the conjunction of the negations of all conditionals is the zero conditional $(0 \mid \vee_i b_i)$.

9.4.5 Corollary

An orthogonal set of conditionals $J = \{(a_i | b_i)\}$ is complete if and only if $\vee_i a_i b_i = \wedge_i b_i$.

9.4.6 Corollary

An orthogonal set of conditionals $J = \{(a_i | b_i)\}$ is complete if and only if $\vee_i (a_i | b_i) = (\wedge_i b_i \mid \vee_i b_i)$.

Thus the disjunction of all the conditionals in a COSC exactly covers the conjunction of all of its conditions. It is this conjunction of the conditions of a COSC that is being orthogonally partitioned.

A special case occurs when $\wedge_i b_i = \vee_i b_i$, which means that all the b_i are equal. Then $J = \{(a_i|b)\}$ reduces to a standard partition in the conditional Boolean algebra $(\mathcal{B}|b)$. If $b = 1$, then the concept reduces to an ordinary partition $\{a_i\}$ of 1 in $\mathcal{B}$.

9.4.7 Expansion Theorem
Let $J = \{(a_i | b_i)\}$ be a set of (pair-wise) orthogonal conditionals and set $b = \vee_i b_i$, then

$$P(\vee_i (a_i | b_i)) = \Sigma_i P(a_i | b_i) P(b_i | b)$$

Proof of Theorem 9.4.7 $P(\vee_i (a_i|b_i)) = P(\vee_i a_i b_i | \vee_i b_i)) = P(\vee_i a_i b_i) / P(b) = \Sigma_i P(a_i b_i) / P(b) = \Sigma_i P(a_i|b_i)P(b_i) / P(b) = \Sigma_i P(a_i|b_i)P(b_i|b)$, using that the $a_i b_i$ are pair-wise disjoint.

9.4.8 Representation Theorem
If $J = \{(a_i | b_i)\}$ is a COSC then for any conditional $(e|f)$,

$$(e|f) = \vee_i (e|f)(a_i|b_i) \text{ if and only if } \vee_i b_i \leq f.$$

Proof of Theorem 9.4.8 $(\vee_i b_i \leq f)$ is necessary because the right hand side of the equation, when expressed as a single conditional, has the condition $\vee_i (f \vee b_i) = f \vee (\vee_i b_i)$, which equals f only if $\vee_i b_i \leq f$. To show sufficiency, suppose $\vee_i b_i \leq f$. So $(\vee_i b_i)f' = 0$. Then

$$\begin{aligned}
\vee_i (e|f)(a_i|b_i) &= (\vee_i (a_i b_i f' \vee b_i' e f \vee a_i b_i e f) | f) \\
&= ((\vee_i a_i b_i)f' \vee (ef)(\vee_i b_i') \vee (ef)(\vee_i (a_i b_i) | f) \\
&= (0 \vee (ef)[(\vee_i b_i') \vee (\vee_i (a_i b_i)] | f) \\
&= (ef[(\wedge_i b_i)' \vee (\wedge_i b_i)] | f) \\
&= (ef[1] | f) = (e|f),
\end{aligned}$$

where $(\vee_i b_i)f' = 0$ has been used and also that $\vee_i (a_i b_i) = \wedge_i b_i$. That completes the proof.

While this logical representation of $(e|f)$ is correct, it is only with further qualification that $P(e|f) = \Sigma_i P(e|f)P(a_i|b_i)$.

9.4.9 Completion Theorem
Every orthogonal system of conditionals can be completed.

Proof of Theorem 9.4.9 Let $J = \{(a_i|b_i)\}$ be a set of pairwise orthogonal conditionals. By the Lemma to the Completeness Characterization Theorem, $\wedge_i b_i - \vee_i a_i b_i = \wedge_i a_i' b_i$ and by the theorem itself, if $\wedge_i a_i' b_i = 0$ then J is already complete. If $\wedge_i a_i' b_i \neq 0$ then $(e|f) = (\wedge_i a_i' b_i \mid \wedge_i b_i)$ is a non-zero conditional that completes J. (e|f) is obviously orthogonal to J. Now if (c|d) is another conditional orthogonal to $J \cup \{(e|f)\}$ then $cd \leq a_i' b_i$ for all i, and also $cd \leq e'f = (\wedge_i a_i' b_i)'(\wedge_i b_i) \leq (\wedge_i a_i' b_i)'$. So $cd \leq (\wedge_i a_i' b_i)(\wedge_i a_i' b_i)' = 0$. So $J \cup \{(\wedge_i a_i' b_i \mid \wedge_i b_i)\}$ is complete. That completes the proof of the theorem.

It is always possible to restrict all the conditionals in a COSC and form a new COSC. If $J = \{(a_i|b_i)\}$ is a COSC, then let (J | f) denote the set of conditionals $\{(a_i \mid b_i f)\}$.

9.4.10 Restriction Theorem
If $J = \{(a_i|b_i)\}$ is a COSC, and f is any proposition, then $(J|f) = \{(a_i \mid b_i f)\}$ is also a COSC.

Proof of Theorem 9.4.10 (J | f) is an orthogonal set because $a_i b_i \leq a_j' b_j$, $i \neq j \Rightarrow a_i b_i f \leq a_j' b_j f$, $i \neq j$. So (J | f) is an orthogonal set of conditionals. (J | f) is complete because if $(c|d) \perp (J \mid f)$, then $cd \leq a_i' b_i f \leq a_i' b_i$ for all i. So by completeness of J, $cd = 0$. So $(c|d) = (0|d)$. Therefore (J | f) is complete.

9.4.11 Second Representation Theorem
If (e|f) is any conditional and $J = \{(a_i|b_i)\}$ is any COSC, then

$$(e|f) = \vee_i (a_i \mid b_i f)(e|f) = \vee_i (e(a_i|b_i) \mid f)$$

Proof of Theorem 9.4.11 By the Restriction Theorem, $(J \mid f) = \{(a_i \mid b_i f)\}$ is a COSC and $\vee_i (b_i f) \leq f$. So by the First Representation Theorem (Section 9.4.8), $(e|f) = \vee_i (a_i \mid b_i f)(e|f)$. Since $(a_i \mid b_i f)(e|f) = (e(a_i|b_i) \mid f)$, the second equality follows. The corresponding probability formulas are in general not true.

9.4.12 Quantum Conditioning

Suppose a physical quantum system and associated measurement apparatus starts out in a condition (state) d, and that we wish to observe the value of the proposition c after some one of a number of mutually exclusive and exhaustive events $\{b_i\}$ occur. In the standard Hilbert linear space interpretation, the "amplitude" or logical representative of getting c via some b_i starting with condition d is

$$\langle c \mid d \rangle = \Sigma_i \langle c \mid b_i \rangle \langle b_i \mid d \rangle$$

and the probability of this is

$$P\langle c \mid d \rangle = |\Sigma_i \langle c \mid b_i \rangle \langle b_i \mid d \rangle|^2$$

(Physically, the square in this formula for the probability of measuring a quantum observable $\langle c \mid d \rangle$ appears to arise from the fact that what is measured is the rate of the energy of a wave at a place and time, which is proportional there and then to the square of the wave's amplitude.)

In the realm of conditionals this formula could be expressed as follows.

9.4.13 Third Representation Theorem

If (c|d) is a conditional and $\{b_i\}$ is a partition of 1, then

$$(c|d) = \vee_i (c \mid b_i d)(b_i d \mid d)$$

and

$$P(c|d) = \Sigma_i P(c \mid b_i d) P(b_i d \mid d).$$

Proof of Theorem 9.4.13 $(c|d) = (c(1) \mid d) = (c(\vee_i b_i) \mid d) = ((\vee_i cb_i) \mid d) = \vee_i (cb_i \mid d) = \vee_i (c \mid b_i d)(b_i d \mid d)$, using that $\{b_i\}$ is exhaustive (i.e. $\vee_i b_i = 1$), and since the members of $\{b_i\}$ are disjoint, $P(c|d) = P(\vee_i (cb_i \mid d)) = \Sigma_i P(cb_i \mid d) = \Sigma_i P(c \mid b_i d) P(b_i d \mid d)$. That completes the proof of the theorem.

Notice also that if c = d, then $P(d|d) = \Sigma_i P(d \mid b_i d) P(b_i d \mid d) = \Sigma_i (1) P(b_i d \mid d) = 1$.

9.4.14 Corollary of Third Representation Theorem
If c is conditionally independent of d given b_i, for all i, that is, if $P(c \mid b_i d) = P(c \mid b_i)$, then

$$P(c \mid d) = \sum_i P(c \mid b_i) P(b_i \mid d).$$

The result follows by substitution and because $P(b_i d \mid d) = P(b_i \mid d)$. In the quantum physical situation described above, it is most often true that starting from an initial condition d, the subsequent occurrence of an intermediate event b_i makes additional subsequent events such as c independent of d. That is, c is conditionally independent of d given b_i.

Note that the conditionals $\{(b_i \mid d)\}$ form a COSC with the property $v_i b_i = 1$. So $v_i(b_i \mid d) = (1 \mid d)$.

In the vernacular of quantum mechanics, $(c \mid b_i)$ is the "projection" of c onto the subspace $(\mathcal{B} \mid b_i)$ and $P(b_i \mid d)$ is the relative probability of the subspace generated by b_i given d.

Notice that in $(\mathcal{B}|\mathcal{B})$ so-called "superpositions" can be expressed as disjunctions of quantum events that have incompatible conditions. For instance, $(a \mid bd) \vee (c \mid b'd)$ exists and has a consistent probability, namely $P(abd \vee cb'd \mid bd \vee b'd) = P(ab \vee cb' \mid d)$. There is no need to believe that this situation is any more mysterious than two conditional alternatives, only one of which applies.

9.4.15 Quantum Vectors and Linear Operator Language
The summation $\sum_i P(c \mid b_i) P(b_i \mid d)$ can be written as the inner product $(\bullet)$ of the two countable vectors

$$\langle P(c \mid b_1), P(c \mid b_2), \ldots P(c \mid b_i), \ldots \rangle \bullet \langle P(b_1 \mid d), P(b_2 \mid d), \ldots P(b_i \mid d), \ldots \rangle$$

or two finite vectors in case $\{(b_i \mid d)\}$ is a finite complete orthogonal set of conditionals (COSC).

An inner product of two n-dimensional real vectors can be interpreted as the product of their lengths times the cosine of the angle between them, and the absolute value of this is the length of one vector times the length of the projection of the 2nd vector onto the first.

The first vector can be considered to be the result of the "projection operator" $[P(\cdot \mid b_1), P(\cdot \mid b_2), \ldots P(\cdot \mid b_i) \ldots]$ applied to any proposition c. It gives the vector of projection (conditional) probabilities of any proposition c on (given) the basis propositions $\{b_i: i=1, 2...\}$.

The second vector is the vector of relative probabilities of the basis propositions given some initial state (proposition) d. The sum over i of the probabilities $P(b_i \mid d)$ is 1. The probability of c is therefore a linear combination of the individual conditional probabilities of c given b_i. If nothing is known about the initial state, then $d = 1$, and so the second vector becomes the vector $\langle P(b_1), P(b_2), ... P(b_i), ... \rangle$ of relative probabilities of the basis propositions, and the equation reduces to

$$P(c) = \sum_i P(c \mid b_i) P(b_i)$$

9.4.16 Quantum Conditioning (Revisited)

In the above, $P(d \mid d)$ represents the probability of a quantum state d that without further measurement perturbation will always be found in state d, and of course that probability is 1. But when, for instance, there is an assumed amplitude from a state d to each member of a complete set of orthogonal states $\{b_i\}$ and then another amplitude back to state d in a subsequent perturbation of the system, the probability $P(\text{not } d \mid d)$ need not be 0 as it must if $P(d \mid d) = 1$. In this double transformation case $P(d' \mid d) = \sum_i P(d' \mid b_i) P(b_i \mid d)$, which need not be zero because the probability distribution (state) has changed. This illustrates why the designers of quantum logic decided to define a "state of the system" as a probability distribution on vectors, vectors built from atoms of complete assignments of values to variables. Every measurement of a variable in general produces a new state (condition) of the probability distribution for the system.

This situation is not yet covered by a theory of conditioning that only *accumulates* conditions and assumes there is no additional measurement disturbance but merely seeks to answer the probability questions of *what else* might be true given that the condition is true.

Nevertheless, even though the measurements are perturbation operators on the probability distribution being measured, the new post-disturbed probability distributions (quantum states) are still defined in $(\mathcal{B} \mid \mathcal{B})$ as sub-spaces $(\mathcal{B} \mid b)$ generated by alternative possible conditions b given to be true in the new state.

9.4.17 Gleason's Theorem
All this brings us to Max Born's [171] assignment of probabilities in Hilbert Space. It was Andrew M. Gleason [172] who proved the uniqueness of Born's probability assignment in Hilbert spaces of 3 or more dimensions. Speaking about additive measures on the collection of subspaces of a Hilbert space, Gleason says:

> It is easy to see that such a measure can be obtained by selecting a vector v and, for each closed subspace A, taking $\mu(A)$ as the square of the norm of the projection of v on A. Positive linear combinations of such measures lead to more examples and, passing to the limit, one finds that, for every positive semi-definite self-adjoint operator T of the trace class,
>
> $$\mu(A) = \text{trace } (TP_A),$$
>
> where P_A denotes the orthogonal projection on A, defines a measure on the closed subspaces. It is the purpose of this paper to show that, in a separable Hilbert space of dimension at least three, whether real or complex, every measure on the closed subspaces is derived in this fashion.

According to Pitowsky [173] Gleason's Theorem can be expressed as follows: Given a probability measure P on a Hilbert space of dimension ≥ 3 there is a self-adjoint, non-negative operator W on $\mathcal{H}$, whose trace is 1, such that P(x) = <**x**,W**x**> for all atoms x ∈ $\mathcal{H}$, where <, > is the inner product, and **x** is a unit vector along x. In particular, if some x_0 ∈ $\mathcal{H}$ satisfies $P(x_0) = 1$ then $P(x) = |<\mathbf{x_0}, \mathbf{x}>|^2$ for all x ∈ $\mathcal{H}$, and that is Born's rule.

9.4.18 Assigning Quantum Probabilities (Explorations)
Starting with a finite set of quantum variables (observables) $\{X_i, i = 1, 2, ... n\}$, suppose that each X_i can take any one of m_i distinct values x_{ij} for $j = 1, 2, ... m_i$. These x_{ij} values could be midpoints of disjoint real intervals where the variables are continuous. Since all matter tends to display periodic orbits and axial spins, quantum variables tend to be associated (in the hidden variable interpretation) with the angular velocities of various orbiting and spinning particles. Each of these individual angular velocities can be characterized by a vector along an axis of rotation whose length is proportional to the magnitude of the angular velocity along that axis.

The energies of these angular momentums are proportional to the squares of their associated angular velocities. In quantum experiments it is these average energies that are measured at a time and place. Furthermore, individual angular momentums can be summed according to vector addition. The energy measured relative to an axis will be periodic and depend on the magnitude of the angular velocity and the difference in angles between the two axes of rotation.

Due to the nature of these variables as angular momentums, the probability of a vector $\langle x_{1j}, x_{2j}, \ldots x_{nj}\rangle$ will be proportional to the sum of the squares of the component values, that is, to the square of the length of the vector.

Denote the event that $(X_i = x_{ij})$ by x_{ij} and form the Boolean algebra $\mathcal{B}$ generated by the collection $\mathcal{B}_a$ of all possible atomic assignments of values to the variables. A typical element of $\mathcal{B}_a$ is a conjunction of n factors, each factor being the event of the ith variable taking a specific one of its values. These n-tuples of complete value assignments for the n variables also form n-dimentional vectors $\langle x_{1j}, x_{2j}, \ldots x_{nj}\rangle$. The number M of atoms (or atomic vectors) formed in this way is the product $m_1 m_2 \ldots m_n$ of the numbers of possible values m_i of each of the n variables. The members of the generated Boolean algebra $\mathcal{B}$ are then the disjunctions (sums) of any subset of the atoms in $\mathcal{B}_a$. These atoms form an orthogonal set of propositions whose various disjunctions ($\vee$) generate the Boolean algebra $\mathcal{B}$.

Now starting with a proposition (vector) v in $\mathcal{B}$ and a closed subspace A of $\mathcal{B}$, the projection of v onto A is the conjunction Av, and the probability of A given proposition v is P(A | v), which is by definition P(Av) / P(v). Identify P(v) as the normalization divisor for the additive measure P(Av) on the subspaces A. Now let $A = \langle A_1, A_2, \ldots A_n\rangle$ where the A_i are the union of values of variable X_i included in subspace A. Similarly, let $v = \langle v_1, v_2, \ldots v_n\rangle$. (Neither v nor A need be atoms.) In any case, $Av = \langle A_1 v_1, A_2 v_2, \ldots A_n v_n\rangle$, and according to Gleason in a Hilbert space, $P(A|v) = P(Av \mid v) = \Sigma_i [P(A_i v_i)]^2 / P(v) = \Sigma_i [P(A_i v_i)]^2 / \Sigma_i [P(v_i)]^2 = |\langle P(Av)\rangle|^2 / |\langle P(v)\rangle|^2$ will be a probability measure on the subspaces A of $\mathcal{B}$. Clearly, this is also sufficient, though not necessary for a probability measure on $(\mathcal{B} \mid v)$.

9.4.19 Hidden Variables Versus Inherent Ambiguity

After Heisenberg's indeterminacy principle limited the theoretical measurement precision of certain pairs of simultaneous measurements like position and velocity, "lack of information" has often been identified with a "state of nature", and the notion of "hidden variables" constituting actual simultaneous positions and velocities, has been seriously questioned, even considered "meaningless" especially in view of mysterious quantum interference and "entanglement phenomena", which imply non-relativistic influences between distant particles or that "particles" have an extended wavelike nature resulting in interference patterns for the probabilities of their possible states and transitions.

However, following David Bohm [117, p77], the probability P of a particle being measured within a volume element dV is given by $P = |\psi|^2 \, dV$, where ψ is Schrödinger's wave function [174], taken as "an objectively real field". Bohm then showed in affect that position-velocity (random) variables exist ("hidden variables") whose joint distributions have averages that agree with any consistent quantum probability distributions as long as faster than light interactions in space are allowed, a mysterious fact of nature more and more acknowledged by quantum physicists.

In this connection, Krennikov [175] has shown how the standard quantum formalism can represent a macroscopic situation involving partial measurement information for which actual simultaneous positions and velocities do exist.

Leonardo Vanni and Roberto Laura [176, 177] demonstrate the potential continuity of quantum properties in time by developing a quantum system that starts with a property p1 at a time t1 and ends with a property p2 at a future time t2, all of which is determined by the "time evolution" (physical mechanics) of the system as given by its Hamiltonian operator H.

Since quantum measurements yield the *average intensities* of identified energy fluctuations, the associated probabilities are statistical averages, not the probabilities of individual events. These underlying (hidden) variables need only have averages that agree with those experimentally derived.

Similarly, Aerts & de Bianchi [178] affirm that the hidden variable interpretation can generate quantum energy distributions as averages of the values of actual variables.

9.5 Deductive Logic and Quantum Operations

We now turn to the question of quantum operations as related to deduction in the realm of conditionals and more generally in the theory of non-monotonic consequence relations.

9.5.1 Deduction and Non-Monotonicity of Conditionals

The sum ideal concept was expanded in [66, p207] and especially in [69] to include deductively closed sets of conditional events with respect to some naturally arising deductive relations on conditionals. If $\leq_x$ is a deductive relation (reflexive and transitive), also called a preorder on the conditionals and J is a set of conditionals, then $H_x(J)$ is defined as the smallest set of conditionals that includes J and which is closed under conjunction and deduction with respect to $\leq_x$. This is the LUB of J with respect to $\leq_x$. Four elementary deductive relations were identified in [69], and earlier [64], and their combinations taken two or three at a time were also observed to be deductive relations. The four elementary deductive relations $\leq_{ap}$, $\leq_{tr}$, $\leq_{nf}$, and $\leq_{ip}$, can be respectively defined by four inequalities for two conditionals (a|b) and (c|d): $b \leq d$, $ab \leq cd$, $a \vee b' \leq c \vee d'$, and $b' \leq d'$.

A remarkable fact rediscovered in the analysis [69, p173] but already present in the writings of Adams [2] is that in determining the implications of a finite set of conditionals, sometimes there is something else required besides the simple conjunction of the members of J. For instance, in determining the deductive consequences of a set $J = \{(a|b), (c|d)\}$ of two conditionals with respect to the deductive relation $\leq_{pm}$, which combines $ab \leq cd$ and $a \vee b' \leq c \vee d'$, the simple conjunction $(a|b) \wedge (c|d)$ is insufficient to deduce all the implications of J. That is, there are conditionals implied with respect to $\leq_{pm}$ by (a|b), which are not implied by $(a|b) \wedge (c|d)$. This flows from the fact that operating with "and" or "or" on conditionals that have non-equivalent premises results in a condtional whose premise is larger than that of either operand.

Therefore, determining the implications of J with respect to $\leq_{pm}$ requires that first all finite conjunctions of members of J be constructed, and then the union of the implications of these conjunctions will embrace all deductive consequences of J with respect to $\leq_{pm}$. So $H_{pm}\{(a|b), (c|d)\} = H_{pm}\{(a|b), (c|d), (a|b)(c|d)\} = H_{pm}(a|b) \cup H_{pm}(c|d) \cup H_{pm}((a|b)(c|d))$. For details see [69, pp173-5].

Recall that $\leq_{pm}$ is the deductive relation for which (a|b) ⊥ (c|d) is true if and only if (a|b) $\leq_{pm}$ (c|d)'. By contrast, with respect to $\leq_\wedge$, the simple conjunction is sufficient for deducing all implications of two or more conditionals. So $H_\wedge\{(a|b),(c|d)\} = H_\wedge((a|b)(c|d))$. See [69, p173] for details.

9.5.4 The Sources of Non-Monotonicity

Engesser and Gabbay [154, p65] attribute the source of non-monotonicity in common sense reasoning to "incomplete information" and also to what they call "self-referential completeness", which they observe applies to quantum mechanics. However, a more direct source of non-monotonicity arises from the operations on conditionals as displayed early by E. Adams [104] and, for instance, in Section 9.2.6 of this chapter.

The importance of making conditions in quantum mechanics explicit cannot be over-stated. The measurement device and configuration must be part of the *condition* for measuring the associated variable and getting a value. The conditioning configuration for measuring some other variable simultaneously may be inconsistent with the first condition and successful measurement. Since the logic of conditional events is already non-monotonic, and since the deductive relations $\leq_x$ and deductively closed sets H_x with respect to them, are also examples of consequence relations in the sense of Engesser and Gabbay, it seems clear that this conditional event algebra and associated deductive theory can be adapted to fully express quantum logic and also satisfy the abstract consequence structures being developed to express the logical content of quantum mechanics.

9.5.5 Propositions and their Models

In his development, Lehmann [155] assumes there is a set $\mathcal{M}$ of models and a "satisfaction" binary relation $\models$ that expresses by $x \models a$ whether a is true in model x. Then the set of models of the propositions in A is $\text{Mod}(A) = \{x \in \mathcal{M}: x \models a \text{ for all } a \in A\}$. Lehmann goes on to define parallel concepts in this domain of extension models allowing expressions of *soundness* and *consistency* and *definability*, the latter being whether a given set is the set of all models for some proposition in $\mathcal{L}$.

In a model, each proposition b is either true or false. This latter statement would not be true if the proposition b were a conditional $(b_1|b_2)$ since then b might be neither true nor false but instead inapplicable. However, it is normally assumed that as applied to an individual model a proposition is either true or false. The conjunction of two propositions, a ∧ b, is true in a model just in case each individually is true in the model.

This well-known "extension" relation between propositions and their sets of models was also used in [63, p 214] to ensure the existence of a logical object (a|b) which could non-trivially hold the "conditional probability, P(a|b), of a given b" of standard probability theory. In fact, P(a|b) was there defined to be P(Mod(a) | Mod(b)).

9.6 Bohm's Model of Quantum Mechanics

David Bohm's "hidden variable" rendition of quantum mechanics [117, p88] provides a way to maintain local properties for particles and express both Heisenberg indeterminacy and quantum entanglement phenomena, although the latter implies faster than light measurement influences between some space-separated particles.

9.6.1 Heisenberg Indeterminacy

To formulate Bohm's indeterminacy framework in the language of Boolean fractions, start with a variable X_k denoting the kth position coordinate of a quantum particle having an average ambient (Schrödinger) field fluctuation δX_k over a time interval Δt. The *intensity* $(\delta X_k)^2$ of this *energy* fluctuation over time Δt is proportional to Δt. So $(\delta X_k)^2 = b\Delta t$, with proportionality constant b. Thus

$$\delta X_k = b^{1/2} (\Delta t)^{1/2}$$

At the same time, the kth component of the average linear momentum p_k of such a particle is proportional to the average change of the velocity (due to measurement disturbance) over that time interval, namely $(\Delta X_k/\Delta t)$. The average particle momentum $p_k = a (\Delta X_k/\Delta t)$, where "a" is another constant of proportionality. Therefore, the average fluctuation δp_k of the momentum is given by

$$\delta p_k = a\, \delta X_k / \Delta t = a\, b^{1/2} (\Delta t)^{1/2} / \Delta t = ab^{1/2}/(\Delta t)^{1/2}$$

Thus the product $\delta X_k \delta p_k$ of the average position fluctuation and the average momentum fluctuation is the constant ab:

$$\delta X_k \delta p_k = ab.$$

Thus there is a maximum possible precision to simultaneous position-velocity measurements of a particle (Heisenberg Indeterminacy) without implying that particles are also waves with infinite extension that collapse when measured at a particular place.

In the language of Boolean fractions this means for any positive epsilon ε, the Boolean fraction $(ab/\varepsilon \leq \delta X_k \mid \delta p_k \leq \varepsilon)$ has probability 1. It is a certainty. That is, given that the momentum error is less than or equal to $\varepsilon > 0$, then the position error is at least ab/ε, where a and b are positive constants. The product (ab) is denoted h, Plank's constant.

Similarly, with the Bohm interpretation and Boolean fractions, there is no need to imagine so-called "superpositions" of two or more events as a mysterious combination of events, none of which actually exist until one is measured. In the algebra of Boolean fractions there exist unions (disjunctions) and conjunctions of two or more Boolean fractions (conditional events) possibly having different conditions. Such an algebraic object is simply a member of the non-Boolean algebra of all Boolean fractions. It generates a non-Boolean sub-algebra. Furthermore, these compound conditionals can be assigned non-trivial probabilities.

9.6.2 Quantum Entanglement

Quantum Entanglement refers to empirically validated phenomena that demonstrate faster-than-light influences when, for instance, the spin direction (+ or -) along different axes of separated particles are measured. The Bohm interpretation of quantum mechanics allows a particle to have actual individual spin along different axes rather than identifying it with an extended wave (the now standard way of thinking). Instead, Schrödinger's interfering waves, closely associated with particles and geometry, are considered real and not exactly identical with the associated particle. As Bohm shows, these waves guide the particles into paths that produce energy interference patterns, and they can also operate to produce the quantum correlations of measurements that constitute entanglement.

It is a good bet that so-called empty space is really a sea of unseen *pre-mass* energies. These pre-mass energies need not be limited to speed of light interactions. Perhaps there is something like a "change of phase" happening as when different gases precipitate into liquid droplets of mass. Such droplets (particles) could have characteristic wavelengths in the ambient energy sea depending on their size. But as they form we should not equate the droplets with their characteristic associated waves in the pre-mass sea. But we should rightly consider them closely related.

In any case, the Bohm interpretation of quantum mechanics is supported by the algebra of Boolean fractions (conditional events) by explaining so-called "super-positions" of two events as simply two alternate events under different possible conditions. The algebra of all conditional events includes objects such as "event A given condition B or event C given condition D". It is not necessary to entertain such an epistemological oddity as that proverbial cat that is "neither alive nor dead" or "both alive and dead". That is a dead end. The Bohm framework instead allows just one condition to be factually true. It does not say that what is presently unknowable implies that it does not exist and that all unknown alternatives are simultaneously both true and false.

9.7 Conclusion

Non-monotonic logics, and in particular, the very expressive 4-operation conditional event algebra ($\mathcal{B}|\mathcal{B}$) developed here, promise to provide a more natural and intuitive formulation of quantum logic than the Hilbert space formulation. The translation of the Hilbert space quantum logic of subspaces into a non-monotonic logical framework would also make all the machinery of the latter theory applicable to the study of quantum mechanics, and vice versa. One way or another, all the quantum concepts, including the standard quantum logical operations in Hilbert Space, which are impossible to express in purely Boolean algebra, can be expressed with this rich, non-distributive, four-operation algebra of conditional events.

References & Bibliography

1. J. C. Abbott (Ed.), *Trends in Lattice Theory*, Van Nostrand Reinhold, 1970.
2. E. W. Adams, *The Logic of Conditionals: An Application of Probability to Deductive Logic*, (Reidel, Boston, 1975).
3. Thomas Bayes, An essay towards solving a problem in the doctrine of chances, *Philos. Trans. Roy. Soc. London LIII*: 370-418 (1763).
4. Thomas Bayes, A demonstration of the second rule in the "Essay towards solving a problem in the doctrine of chances" *Philos. Trans. Roy. Soc. London LIV*: 296-325 (1764).
5. Evert W. Beth, *The Foundations of Mathematics*, North-Holland, Amsterdam, 1959.
6. Garrett Birkhoff, *Lattice Theory*, Amer. Math. Soc. Colloq. Publ., 1948.
7. George Boole, *An Investigation of the Laws of Thought on which are Founded the Mathematical Theories of Logic and Probabilities*, 2nd ed., Open Court, 1940; 1st ed., Macmillan, 1854.
8. Philip G. Calabrese, The Menger algebras of two-place functions in the 2-valued logic, *Notre Dame J. Formal Logic VIII*(4):333-340 (Oct. 1966).
9. Philip G. Calabrese, The probability that p implies q (preliminary report), *Notices Amer. Math. Soc.* **22(3)**: A430-A431 (Apr. 1975).
10. Philip G. Calabrese, The probability of the triangle inequality in probabilistic metric spaces, *Aequationes Math.* 18:187-205 (1978).
11. Rudolf Carnap, *Logical Foundations Of Probability, 2nd ed.*, Univ. of Chicago Press, 1960; 1st ed., 1950.
12. Rudolf Carnap, *Meaning and Necessity: A Study in Semantics and Modal Logic, 2nd ed.*, Univ. of Chicago Press, 1958.
13. Rudolf Carnap and Richard C. Jeffrey, *Studies in Inductive Logic and Probability*, Univ. of California Press, 1971.
14. C. C. Chang and H. J. Keisler, *Model Theory*, North-Holland, 1973; 2nd ed., 1977.
15. Zolton Domotor, Probabilistic Relation Structures and Their Application, *Technical Report No. 144*, Inst, for Mathematical Studies in the Social Sciences, Stanford Univ., Stanford, Calif., 1969.
16. H. Gaifman, Concerning measures in first order calculi, *Israel J. Math.* 2:1-18 (1964).

17. K. Gödel, Die Vollstandigkeit der Axiome des logischen Funktionenkalkuls, *Monatsh. Math. Phys. 37*:349-360 (1930).
18. I. R. Goodman and H. T. Nguyen, *Uncertainty Models for Knowledge-Based Systems*, North-Holland, 1985.
19. Madan M. Gupta (Ed.), *Advances in Fuzzy Set Theory and Applications*, North-Holland, 1979.
20. T. Hailperin, *Boole's Logic and Probability*, 2nd ed., North-Holland, Amsterdam, 1986. (1st ed. 1976)
21. T. Hailperin, *Probability logic*, Notre Dame J. Formal Logic *25(3)*:198-212 (July 1984).
22. Paul R. Halmos, *Measure Theory*, Van Nostrand, 1950.
23. Paul R. Halmos, *Algebraic Logic*, Chelsea, New York, 1962.
24. Jean Van Heyenoort, *From Frege to Godel (a Source Book In Mathematical Logic 1879-1931)*, Harvard UP, 1967.
25. Kaarlo Hintikki and Patrick Suppes (Eds.), *Aspects of Inductive Logic*, North Holland, Amsterdam, 1966.
26. Nathan Jacobson, *Lectures in Abstract Algebra*, Van Nostrand, 1951.
27. A. N. Kolmogorov, *Grundbegriffe der Wahrscheinlichkeitsrechnung*, Berlin, 1933.
28. A. N. Kolmogorov, *Foundations of the Theory of Probability*, Chelsea, 1956.
29. Bart Kosko, Fuzzy cognitive maps, *Internal. J. Man-Machine Stud. 24*:65-67 (Jan. 1986).
30. C. I. Lewis and C. H. Langford, *Symbolic Logic, 2nd ed.*, Dover, 1959; 1st ed., Century, New York, 1932.
31. J. Lukasiewicz, On three-valued logic, *Ruch Filozoficzny* **5**:169-171 (1920).
32. E. Lusk, Priority assignment: A conditioned sets approach, in *Fuzzy Sets and Systems*, North-Holland, Amsterdam, 1984.
33. Stefan Mazurkiewicz, Zur Axiomatik der Wahrscheinlichkeitslehre, *Comptes rendus des séances de la Société des sciences et des Lettres de Varsovie, Classe III, Vol* **25**, (1932) 1-4.
34. Stefan Mazurkiewicz, Über die Grundlagen der Wahrscheinlichkeitsrechnung, *Monatshefte für Mathematik und Physik*, Vol **41** (1934) 343-352.
35. Stefan Mazurkiewicz, Podstawy Rachunku Prawdopodobiehstwa (*The Foundations of the Calculus of Probability*), Monografie Matematyczne, *Vol. 32*, Pahstwowe Wydawnietwo Naukowe, Warszawa, 1956.

36. Donald Nute, *Topics In Conditional Logic*, Reidel, 1980.
37. Howard Pospesel, Arguments: *Deductive Logic Exercises*, Prentice-Hall, 1971, p27, #78.
38. Emil L. Post, *Introduction to a general theory of elementary propositions*, Amer. J. Math.
39. Henri Prade, A computational approach to approximate and plausible reasoning with applications to expert systems, *IEEE Trans. Pattern Analysis and Machine Intelligence PAMI-7*, No. 3 (May 1985).
40. Arthur N. Prior, *Time And Modality*, Oxford U.P., 1957.
41. Arthur N. Prior, *Past, Present and Future*, Oxford U.P., 1967.
42. Willard V. Quine, *Methods of Logic, 4th ed.*, Harvard U.P., 1982.
43. Alfred Renyi, *Foundations of Probability Theory*, Holden Day, San Francisco, 1970.
44. Nicholas Rescher, A probabilistic approach to modal logic, *Acta Philos. Fenn.* 16:215-226 (1963).
45. Nicholas Rescher, An intuitive interpretation of systems of four-valued logic, *Notre Dame J. Formal Logic* 6:154-156 (1965).
46. Nicholas Rescher, *Many-Valued Logic*, McGraw-Hill, 1969.
47. Nicholas Rescher and Alasdair Urquhart, *Temporal Logic*, Springer, New York, 1971.
48. Nicholas Rescher, *Topics in Philosophical Logic*, Reidel, Dordrecht, 1968.
49. Paul C. Rosenbloom, *The Elements of Mathematical Logic*, Dover, 1950.
50. Bertrand A. W. Russell and Alfred N. Whitehead, *Principia Mathematica*, Cambridge U.P., 1913, 3 vols.
51. D. Scott and Paul Krauss, Assigning probabilities to logical formulas, in *Aspects of Inductive Logic* (J. Hintikki and P. Suppes, Eds. 1966, pp. 219-264.
52. G. Shafer, *A Mathematical Theory of Evidence*, Princeton UP, Princeton, NJ, 1976.
53. Patrick Suppes et al, *New foundations of objective probability: Axioms for propensities, in Logic, Methodology and Philosophy of Science*, Bucharest, 1971 (P. Suppes, L. Henkin, Gr. C. Moisil, and A. Joja, Eds.) North-Holland, Amsterdam, 1973.
54. M. H. Stone, The theory of representations for Boolean algebras, *Trans. Amer. Math. Soc.* **40**:37-111 (1936).
55. Alfred Tarski, Wahrscheinlichkeitslehre und mehrwertige Logik, *Erkenntnis* 5:174-175 (1935-36).

56. Alfred Tarski, *Logic, Semantics, Metamathematics — Papers from 1923 to 1938*, Oxford, 1956.
57. John Venn, *The Logic Of Chance*, London, 1888.
58. L. A. Zadeh, Fuzzy sets, *Inform, and Control* **8**:338-353 (1965).
59. Bruno De Finetti, *Probability, Induction and Statistics*, (Wiley, New York, 1972).
60. G. Schay, An Algebra of Conditional Events, *Journal of Mathematical Analysis and Applications* **24** (1968) 334-344.
61. D. Lewis, Probabilities of Conditionals and Conditional Probabilities, *The Philos. Rev.* **85 (3)** (1976) 297-315.
62. N. J. Nilsson, Probabilistic Logic, *Artificial Intelligence* **28 (1)** (1986) 71-87.
63. P. G. Calabrese, An Algebraic Synthesis of the Foundations of Logic and Probability, *Information Sciences* **42** (1987) 187-237.
64. P. G. Calabrese, Reasoning with Uncertainty Using Conditional Logic and Probability, in: *Proc First International Symposium on Uncertainty Modeling and Analysis*, IEEE Computer Society, (1990) 682-688.
65. P. G. Calabrese, Deduction and inference using conditional logic and probability, Chap. 2 in: I. R. Goodman, M.M. Gupta, H.T. Nguyen and G.S. Rogers, eds., *Conditional Logic in Expert Systems*, (North-Holland, Amsterdam, 1991) 71-100.
66. P. G. Calabrese, Conditional events: doing for logic and probability what fractions do for integer arithmetic (Invited talk & paper) in: *International Conference The Notion of Event in Probabilistic Epistemology*, Publicazione n.4, Università Degli Studi di Trieste, Departimento di Matematica Applicata, Alle Scienze Economiche e Statistiche ed Attuariali, Trieste, May 27-29, 1996, 175-212 http://www.stormingmedia.us/22/2275/A227503.html
67. P. G. Calabrese, A theory of conditional information with applications, In *Special Issue on Conditional Event Algebra, IEEE Transactions on Systems, Man, and Cybernetics*, Vol. 24, No. 12 (1994) 1676-1684.
68. P. G. Calabrese, An extension of the fundamental theorem of Boolean algebra to conditional propositions, Part I (1-18) of "Conditional event algebras and conditional probability logics" by P. G. Calabrese and I. R. Goodman, *Probabilistic Methods in Expert Systems*, Romano Scozzafava ed., Proceedings of the International Workshop, Rome, Italy, Societá Italiana di Statistica, (14-15 Oct. 1993) 1-35.

69. P. G. Calabrese, Deduction with uncertain conditionals, *Information Sciences* **147** (2002) 143-191
http://dx.doi.org/10.1016/S0020-0255(02)00262-1
70. P. G. Calabrese, Operating on functions with variable domains, *J. Philosophical Logic*, (Feb. 2003), **32**(1), 1-18.
http://www.springerlink.com/content/t26j823487w5t0p/
71. P. G. Calabrese, Reflections on logic & probability in the context of conditionals, (Invited Paper) in *Conditionals, Information, and Inference*, G. Kern-Isberner, W. Rödder, and F. Kulmann (Eds.) Revised Selected Papers of Workshop on Conditionals (WCII 2002), Springer-Verlag *Lecture Notes in Artificial Intelligence* (LNAI) 3301 (2005) 12–37.
http://www.springerlink.com/content/mdqd771t8rtq42uj/
72. P. G Calabrese, Toward a more natural expression of quantum logic with Boolean fractions, *J. Philosophical Logic* **34** (2005), no. 4, 363–401.
http://www.springerlink.com/content/l8j015564585h568/ First draft available at http://arXiv.org/abs/quant-ph/0305009
73. P. G. Calabrese, The logic of quantum measurement in terms of conditional events", *Oxford University Press Logic Journal of the IGPL (2006)* **14(3)**: 435-455.
http://dx.doi.org/10.1093/jigpal/jzl018/
74. I. R. Goodman, Three-valued logics and Conditional Event Algebras, *Proc First International Symposium on Uncertainty Modeling and Analysis* (IEEE Computer Society, 1990) 31-37.
75. I. R. Goodman, H. T. Nguyen and E. A. Walker, *Conditional Inference and Logic for Intelligent Systems: a Theory of Measure-free Conditioning*, North-Holland, Amster-dam, 1991).
76. I. R. Goodman, Algebraic and Probabilistic Bases for Fuzzy Sets and the Development of Fuzzy Conditioning, in: I. R. Goodman, M. M. Gupta, H. T. Nguyen and G. S. Rogers, eds., *Conditional Logic in Expert Systems* (North-Holland, Amsterdam, 1991) 1- 69.
77. I. R. Goodman, Evaluation of Combinations of Conditioned Information: A History, *Information Sciences* **57-58** (1991) 79-110.
78. H. T. Nguyen and G. S. Rogers, Conditioning Operators in a Logic of Conditionals, in: I. R. Goodman, M. M. Gupta, H. T. Nguyen and G. S. Rogers, eds., *Conditional Logic in Expert Systems* (North-Holland, Amsterdam, 1991) 159-179.

79. I. R. Goodman, M. M. Gupta, H. T. Nguyen and G. S. Rogers Eds. *Conditional Logic in Expert Systems* (North-Holland, Amsterdam, 1991)
80. H. T. Nguyen, Algebraic Structures of Conditioning Reasoning, in: *Proc First International Symposium on Uncertainty Modeling and Analysis* (IEEE Computer Society, 1990) 564-566.
81. J. Pearl, *Probabilistic Reasoning in Intelligent Systems: Networks of Plausible Inference*, (Morgan Kaufman, San Mateo, 1988).
82. E. A. Walker, A Simple Look at Conditional Events, in: I. R. Goodman, M. M. Gupta, H. T. Nguyen and G. S. Rogers, eds., *Conditional Logic in Expert Systems* (North-Holland, Amsterdam, 1991) 101-114.
83. D. Dubois and H. Prade, The logical View of Conditioning and its Application to Possibility and Evidence Theories, *International Journal of Approximate Reasoning* **4** (1990) 23-46.
84. D. Dubois and H. Prade, Conditioning, non-monotonic logic, and non-standard uncertainty models, in: I. R. Goodman, M.M. Gupta, H.T. Nguyen and G.S. Rogers, Eds. *Conditional Logic in Expert Systems*, North-Holland, Amsterdam, 1991, 115-158.
85. T. Hailperin, *Sentential Probability Logic*, Lehigh University Press, 1996.
86. P. Chrzastowski-Wachtel, A. Hoffmann, J. Tyszkiewicz, A. Ramer, Definability of Connectives in Conditional Event Algebras of Schay-Adams-Calabrese and Goodman-Nguyen-Walker, *Information Processing Letters 01/2001*; **79**:155-160.
87. J. Tyszkiewicz, A. Hoffmann, & Arthur Ramer, Embedding Conditional Event Algebras Into Temporal Calculus of Conditionals, February 1, 2008, arXiv:cs/0110003v1 [cs.AI] 1 Oct 2001.
88. E. W. Walker, Stone algebras conditional events and three valued logic, *IEEE Transactions of Systems, Man and Cybernetics* **24 (12)** (1994) 1699-1707.
89. B. Sobocinski, Axiomatization of a partical system of the three-valued calculus of propositions, *Journal of Computing Systems* **1** (1952) 23-25.
90. B. De Finetti, La logique de la probabilite, in *Actes du Congres International de Philosophie Scientifique* Paris: Hermann Editeurs, **IV**, (1936) 1-8.
91. P. Suppes and M. Zanotti, Z. Wahrscheinlichkeitstherie verw. *Gebiete* **60** (1982) 163-169. (Also in [92].)

92. P. Suppes and M. Zanotti, *Foundations of Probability with Applications, Selected Papers 1974-1995*, Cambridge University Press (1996).
93. A. Papoulis, *Probability, Random Variables, and Stochastic Processes*, McGraw-Hill (1965).
94. E. W. Adams, Probability and the Logic of Conditionals, in: *Aspects of Inductive Logic*, P. Suppes & J. Hintikka, eds., Amsterdam: North-Holland (1966) 265-316.
95. E. W. Adams, On the logic of conditionals, *Inquiry* 8 (1965) 166-197.
96. D. Dubois, and H. Prade, "Conditional Objects as Nonmonotonic Consequence Relationships", *IEEE Transactions on Systems, Man, and Cybernetics*, Vol. **24**, No. 12 (1994) 1724-1740.
97. D. Dubois, H. Prade, Conditional Objects and Non-Monotonic Reasoning, *Proc of the Second International Conference on Principles of Knowledge Representation and Reasoning*, J. Allen, R. Fikes and E. Sandewall Eds. Cambridge, Massachusetts (April 1991) 22-25.
98. L. Zadeh, The Role of Fuzzy Logic in the Management of Uncertainty in Expert Systems, *J. Fuzzy Sets and Systems*, **11** (1984), 199-227.
99. R. Laura & L. Vanni, Time Translation of Quantum Properties. *Found. Phys. 39*, 160–173, (2008)
100. L. Vanni & R. Laura, The Logic of Quantum Measurements, *Int J Theor Phys (2013)* **52**:2386–2394, DOI 10.1007/s10773-013-1522-6
101. J. Tyszkiewicz, A. Hoffmann, & Arthur Ramer, Embedding Conditional Event Algebras Into Temporal Calculus of Conditionals, February 1, 2008, arXiv:cs/0110003v1 [cs.AI] 1 Oct 2001
102. W. V. O. Quine, 'Two Dogmas of Empiricism' *from a Logical Point of View*, 2nd edn. Cambridge, MA: Harvard University Press, 1961.
103. H. Putnam, Is Logic Empirical?, *Boston Studies in the Philosophy of Science, Vol 5, Eds*. Robert S. Cohen and Marx W. Wartofsky (Dordrecht: D. Reidel, 1968), pp. 216-241. Repr. as The Logic of Quantum Mechanics in *Mathematics, Matter and Method* (1975), pp. 174-197.
104. H. Putnam, How to think quantum-logically, in: *Logic and Probability in Quantum Mechanics*, P. Suppes ed., D. Reidel, 1976, 47-53.

105. P. Suppes and M. Zanotti, Stochastic incompleteness of quantum mechanics, *Synthese* **29** 1974, 311-330; also in [106] pp 303-322 and in [92] pp 67-82.
106. *Logic and Probability in Quantum Mechanics*, P. Suppes ed., D. Reidel, 1976
107. P. Suppes, Probabilistic causality in quantum mechanics, *J. Statistical Planning and Inference* **25**, 1990, 293-302
108. A. Wilce, Quantum Logic and Probability Theory, *The Stanford Encyclopedia of Philosophy* (Spring 2003 Edition), Edward N. Zalta (ed.), http://plato.stanford.edu/archives/spr2003/entries/qt-quantlog/
109. M. Planck, On an Improvement of Wien's Equation for the Spectrum, *Deutsch. Phys. Gesell. Verh.*, *2*, 1900, 202-204 and On the theory of the energy distribution law of the normal spectrum *Deutsch. Phys. Gesell. Verh.*, *2*, 1900, 237-245.
http://hermes.ffn.ub.es/luisnavarro/clasicos_2.htm
www.**law**ebdefisica.com/arts/**distributionlaw**.pdf
110. A. Einstein, Zur Theorie der Lichterzeugung und Lichtabsorption / On the Theory of Light Production and Light Absorption, *Annalen der Physik*, Leipzig 20 (1906) 199.
111. L. de Broglie, Radiations — *Ondes et Quanta*/Radiation – *Waves and Quanta*, *Comptes Rendus, Vol. 177* (1923) 507-510.
112. E. Schrödinger, Quantisierung als Eigenwertproblem Erste Mitteilung, *Annalen der Physik*, **79**, 1926, 361-489.
113. P. Dirac, The physical interpretation of the quantum dynamics, *Proc. Royal Soc. of London A 113* (1926) 621-641.
114. J. von Neumann, *Mathematical Foundations of Quantum Mechanics*, Translated from German (1932) by R.T. Beyer, Princeton Univ. Press, 1955.
115. G. Birkhoff & J. von Neumann, The logic of quantum mechanics, *Annals of Mathematics* 37, 1936, 823-834.
116. W. Heisenberg, *The Physical Principles of the Quantum Theory*, Dover, 1930.
117. D. Bohm, *Wholeness and the Implicate Order*, Ark Paperbacks, 1983. (First published in 1980.)
118. T. Fine, Towards a revised probabilistic basis for quantum mechanics, in *Logic and probability in quantum mechanics*, P. Suppes ed., D. Reidel, 1976, 179-193.
119. J. S. Bell, On the Einstein-Podolsky-Rosen paradox, *Physics 1*, 1964, 195-200.

120. J. S. Bell, On the problem of hidden variables in quantum mechanics, *Rev. Mod. Phys.* **38**, 1966, 447-452.
121. J. S. Bell, *Speakable and Unspeakable in Quantum Mechanics*, Cambridge University Press, 1987.
122. S. Goldstein, Bohmian mechanics and the quantum revolution, *Synthese* **106**, Feb. 1996. URL = http://arxiv.org/abs/quant-ph/9512027
123. S. Goldstein, Quantum philosophy: The flight from reason in science, in *The Flight from Science and Reason*, edited by P. Gross, N. Levitt, and M. W. Lewis, *Annals of the New York Academy of Sciences*, 1996. URL = http://arxiv.org/abs/quant-ph//9601007
124. S. Goldstein, "Bohmian mechanics", *The Stanford Encyclopedia of Philosophy* (Winter 2002 Edition), E. N. Zalta ed., URL = http://plato.stanford.edu/archives/win2002/entries/qm-bohm
125. W. Rödder, Conditional logic and the Principle of Entropy, *Artificial Intelligence* **117** (Feb. 2000), 83-106
126. C. Norris, Putnam on Quantum Theory and Three-Valued Logic: Is It (Realistically) an Option? *Journal of Critical Realism (incorporating Alethia)* **5.1** (May 2002): 39-50.
URL=http://www.journalofcriticalrealism.org/archive/JCR(A)v4n2_norris23.pdf
127. L. E. Ballentine, Probability theory in quantum mechanics, *Am. J. Phys.* **54** (10) October 1986.
128. R. P. Feynman, R. B. Leighton & M. Sands, *Lectures on Physics Vol. III (Quantum Mechanics)*, 1965.
129. B. O. Koopman, Quantum theory and the foundations of probability, in *Applied Probability*, L. A. MacColl, ed., McGraw-Hill, 1955, 97-102.
130. A. Einstein, B. Podolsky, & N. Rosen, Can quantum-mechanical description of physical reality be considered complete?, *Physical Review* **47**, 1935, 777-780.
131. A. Fine, Probability and the interpretation of quantum mechanics, *British J. for the Philosophy of Science* **24**, 1973, 1-37.
132. A. Fine, Hidden variables, joint probability, and the Bell inequalities, *Phys. Rev. Lett.* **48**, No.5, 1982, 291-295.
133. A. N. Kolmogorov, *Foundations of the Theory of Probability*, Chelsea (1956); (1st edition: *Grundbegriffe der Wahrscheinlichkeits-rechnung*, Berlin, 1933.)

134. A. Khrennikov, 'Quantum probabilities' as context depending probabilities, 13 Jun 2001, 1-6. URL = http://arXiv.org/abs/quant-ph/0106073
135. A. Khrennikov, "Contextual viewpoint to quantum stochastics", Dec. 2001, 1-13. URL = http://arXiv.org/abs/hep-th/0112076
136. A. Khrennikov, "Växjö interpretation of quantum mechanics", Feb. 2002, 1-11. URL = http://arXiv.org/abs/quant-ph/0202107
137. A. Khrennikov, "Local realist (but contextual) derivation of the EPR-Bohm correlations", Nov. 2002, 1-17. URL = http://arXiv.org/abs/quant-ph/0211073
138. I. Pitowsky, "Betting on outcomes of measurements: A Bayesian theory of quantum probability", Aug. 2002. URL = http://arxiv.org/abs/quant-ph//0208121
139. I. Pitowsky, *Quantum Probability – Quantum Logic*, Lecture Notes in Physics 321, Springer, 1989.
140. C. A. Fuchs, Quantum mechanics as quantum information, 8 May 2002, 1-59. URL = http://arxiv.org/abs/quant-ph/0205039
141. A. Landé, Why the world is a quantum world, in: *Logic and probability in quantum mechanics*, P. Suppes ed., D. Reidel, 1976, 433-444.
142. P. Busch, Is the quantum state (an) observable? In: *Experimental Metaphysics – quantum mechanical studies in honor of Abner Shimony* R. S. Cohen and J. Stachel eds., D. Reidel, Dordrecht, 1996, 1-12. URL = http://arxiv.org/abs/quant-ph9604014
143. P. Busch, "Just how final are today's quantum structures?" Springer Forum: Quantum Structures – Physical, Mathematical and Epistemological Problems, J. Liptovsky, 1 September 1998. URL = http://arxiv.org/abs/quant-ph/0103139
144. V. S. Varadarajan, *Geometry of Quantum Theory*, Vols. 1 & 2, D. Van Nostrand, 1968.
145. J. M. Jauch & C. Piron, On the structure of quantal proposition systems, *Helvetika Physica Acta 42*, 1969, 842-848.
146. J. M. Jauch, The quantum probability calculus, in: *Logic and probability in quantum mechanics*, P. Suppes ed., D. Reidel, 1976, 123-146.
147. B. Coecke, D. Moore, & A. Wilce, Operational quantum logic: an overview, introductory chapter of *Current research in operational quantum logic: algebras, categories, languages, Fundamental theories of physics series*, Kluwer Academic Publishers, 2000. URL = http://arxiv.org/abs/quant-ph/0008019

148. C. Piron, *Foundations of Quantum Physics*, W. A. Benjamin, Inc., 1976.
149. B. Coecke, Disjunctive quantum logic in dynamic perspective, 2002. http://arXiv.org/abs/math/0011209
150. U. Sasaki, Lattices of projections in AW^*-algebras, *J. of Science of Hiroshima University A* **19**, 1955, 1-30.
151. Y. Delmas-Rigoutsos, A double deduction system for quantum logic based on natural deduction, *J. Philosophical Logic* **26**, No. 1, 1997, 57-67.
152. D. Aerts, E. D'Hondt, & L. Gabora, Why the disjunction in quantum logic is not classical, *Foundations of Physics*, *Vol. 30*, Issue 10, 2000. URL = http://arXiv.org/abs/quant-ph/0007041
153. E. W. Adams, *A Primer of Probability Logic*, CSLI Publications, Stanford, CA, 1998.
154. K. Engesser & D. M. Gabbay, Quantum logic, Hilbert space, revision theory, *Artificial Intelligence*, **136(1)**, March 2002, 61-100.
155. D. Lehmann, Connectives in Quantum and Other Cumulative Logics, 2 Aug. 2002, URL = http://arxiv.org/abs/cs.AI/0205079
156. "Is logic empirical?", Reviews of papers by H. Putnam and Michael Dummett, 2004. *Wikipedia, the free encyclopedia*, URL = http://www.unipedia.info/Is_logic_empirical.html
157. W. Rödder, C. H. Meyer, Coherent Knowledge Processing at Maximum Entropy by SPIRIT, *Proc Twelfth Conf On Uncertainty in Artificial Intelligence*, Portland, Oregon, USA, (1996) 470-476.
158. M. Smyth, Power Domains, *Journal of Computer Sciences*, **16** (1978) 23-36.
159. G. Winskel, On Powerdomains and Modality, *Theoretical Computer Science*, **36** (1985)127-137.
160. S. Abramsky, Domain Theory in Logical Form, *Annals of Pure and Applied Logic*, 1988.
161. I. R. Goodman & H.T. Nguyen, Mathematical Foundations of Conditionals and Their Probabilistic Assignments, *Journal of Uncertainty, Fuzziness and Knowledge-Based Systems* 3 (3) (1995) 247-339.
162. E. W. Adams, On the Logic of High Probability, *Journal of Philosophical Logic* **15** (1986) 225-279.
163. G. M. Hardegree, The conditional in quantum logic, in *Logic and Probability in Quantum Mechanics*, P. Suppes ed., D. Reidel, 1976, 55-72.

164. J. N. Hooker, A Quantitative Approach to Logical Inference, *Decision Support Systems* (4) (1988) 45-69.
165. J. E. Shore, "Axiomatic Derivation of the Principle of Maximum Entropy and the Principle of Minimum Cross-Entropy", IEEE Transactions on Information Theory, IT-26, 1 (Jan. 1980), 26-37.
166. E.T. Jaynes, "Concentration of Distributions at Entropy Maxima", reprinted in R. D. Rosenkrantz (ed.), E.T. Jaynes: Papers on Probability, Statistics and Statistical Physics, Reidel, Dordrecht, 1983.
167. S. Amari, Differential-Geometrical Methods in Statistics, Springer-Verlag, 1985.
168. A. Caticha, "Maximum Entropy, Fluctuations and Priors", Max-Ent 2000, the 20th Interna-tional Workshop on Bayesian Inference and Maximum Entropy Methods, July 8-13, 2000, Gif-sur-Yvette, France.
169. S. Amari & T.S. Han, "Statistical Inference Under Multiterminal Rate Restrictions: A Differential Geometric Approach", Information Theory, IEEE Transactions on, Volume: 35 Issue: 2, (March 1989) 217–227.
170. C. R. Rao, "Information and the Accuracy Attainable in the Estimation of Statistical Pa-rameters", Bulletin Calcutta Mathematical Society, Vol. 35, (1945) 199-210.
171. M. Born, "Zur Quantenmechanik der Stoßvorgänge". *Zeitschrift für Physik* **37** *(12) (1926) 863–867*. Bibcode:*1926ZPhy...37..863B*. doi:*10.1007/BF01397477*.
172. A. M. Gleason, "Measures on the closed subspaces of a Hilbert space" *Indiana University Mathematics Journal* **6** (1957) 885–893. doi:10.1512/iumj.1957.6.56050.
173. Itamar Pitowsky, "Quantum Mechanics as a Theory of Probability", February 1, 2008, arXiv:quant-ph/0510095.
174. E. Schrödinger, "An Undulatory Theory of the Mechanics of Atoms and Molecules" (PDF). Physical Review **28** (6) (1926) 1049–1070. Bibcode:*1926PhRv...28.1049S*. doi:*10.1103/PhysRev.28.1049*.
175. A. Khrennikov, "Quantum-like representation of macroscopic configurations, in *Quantum Interaction*, (2009) 44--58, Lecture Notes in Comput. Sci., 5494, Springer, Berlin.
176. R. Laura & L. Vanni, "Time translation of quantum properties" Found. Phys. 39 (2009), no. 2, 160–173. MR2475712 (2009k:81011).

177. R. Laura & L. Vanni, "The Logic of Quantum Measurements", Int J Theor Phys (2013) 52:2386–2394 DOI 10.1007/s10773-013-1522-6.
178. D. Aerts and M. Sassoli de Bianchi, in The extended Bloch representation of quantum mechanics and the hidden-measurement solution to the measurement problem. Annals of Physics 351 (2014) 975–1025 (Open Access)
179. E. T. Jaynes, "Probability in Quantum Theory", in *Complexity, Entropy and the physics of Information*, W. H. Zurek, Ed, Addison Wesley Pub. Co., Reading, MA (1990), p. 4.
180. E. T. Jaynes, "Information Theory and Statistical Mechanics", *The Physical Review*, Vol. 106, No. 4, 620-630, May 15, 1957.
181. J. Lukasiewicz & A. Tarski, 'Untersuchungen über den Aussagenkalkül' (Investigations into the Sentential Calculus), in Comptes Rendus des séances et de la Société des Sciences et des Lettres de Varsovie, vol. 23, 1930, cl. iii, pp. 30-50.
182. Aristotle, *De Interpretatione*, Translated by L. Ackrill, ed. *A New Aristotle Reader*, Princeton University Press, 1987 as quoted in "Future Contingent Propositions and the Law of Excluded Middle in Aristotle and Some Philosophers", Richard G. Howe, Ph.D., 2006.
http://richardghowe.com/index_htm_files/FutureContingencies.pdf

Index

6-sided die, 22, 63, 81, 84, 173, 174

Abbott, J. C., 253
Abramsky, S., 263
absent-minded coffee drinker, 154, 156
Adams, Ernest W., 2, 7, 22, 24-25, 33, 46, 62, 65, 119, 136, 199, 231, 247-248, 253, 259, 263
addition, 27, 46, 75, 76-78, 88, 92, 93, 101, 161, 185, 198, 245
additional deductive information, 142, 164
additive law, 196, 208, 228
Aerts, D., 222, 246, 263, 265
algebra of conditionals, 24, 31, 33, 37-38, 82, 121, 189, 196, 202, 206, 216, 219, 223, 226, 236, 237
algebraic logic, v, 5-6, 43, 45, 95, 98, 254
all times, 9, 180-181, 186
Amari, S., 173-174, 264
angular momentum, 245
angular velocity, 244-245
antecedent, 54, 145, 200
applicability, 23, 42, 125, 144, 158, 159, 176, 190, 192-193, 213, 215-216, 220, 226, 233, 235
Aristotle, 4, 15, 265
atomic vectors, 245
atoms, 111-112, 115, 158, 184, 243-245, 264
axiom system, 101, 177

Ballentine, L. E., 223, 225, 228, 261
Bayes, Thomas, 108, 253
Bayesian, iii, iv, 33, 35, 171-173, 191, 262, 264
Bell, J. S., 190, 193, 225, 228, 260-261
Beth, Evert W., 7, 253
Birkhoff, Garrett, 7, 190-191, 227, 253, 260

Bohm, David, 190, 225, 227, 246, 249-251, 260
Boole, George, 1, 7, 10, 19, 22, 24, 25, 46, 108, 195, 253, 254
Boolean algebra, 11-13, 18-20, 24-25, 28-29, 33, 43-44, 46-49, 52-56, 58, 66, 69-70, 111, 121, 138, 148, 158, 162-163, 175-176, 190, 200-202, 208, 212-213, 217-218, 222-223, 225, 228-229, 231, 233, 235, 239, 245, 251, 255-256
Boolean components, 65, 110, 127, 175, 176
Boolean deduction, 39, 82, 120, 125, 128, 148, 149, 153, 155, 218
Boolean extension property, 119, 120, 126, 131
Boolean fraction, v, 108-111, 113, 175, 184, 189, 191-192, 195-196, 200-201, 204-205, 209, 212, 220, 225, 227, 229-230, 249-251, 257
Boolean function, 2, 12, 13, 16-17, 58, 62, 69-73
Boolean logic, iii, 1-2, 5, 8, 11-12, 34, 45, 97, 105, 108, 110-111, 114, 149, 154, 192, 195, 221, 227, 229
Boolean proposition, 5, 22, 41, 51, 54, 69, 70, 78, 100, 109, 110-111, 121, 143, 151, 200, 204, 206, 221, 223
Boolean sub-algebra, 120, 190, 196, 216, 218, 226, 228, 231, 236
Borel set, 85
Born, Max, 244, 264
Busch, P., 208-209, 262

Calabrese, Philip G, i, iv, 2, 4, 16, 46, 47, 54, 64, 81, 87, 119, 126, 127, 143, 184, 199-200, 226, 253, 256-257
canonical development, 115
Carnap, Rudolf, 7, 253
Caticha, A., 173, 175, 264

certainty, 22, 119, 123, 127, 171, 222, 250
Chang, C. C., 7, 43, 100, 253
Chrzastowski-Wachtel, P., 258
Coecke, B., 219, 262-263
combined deductive relation, 130, 134, 164
commutative law, 96, 142
compatibility, 189, 191, 195, 221
compatible conditionals, 196
complement, 1, 18, 21, 22, 48, 77-78, 111, 151-152, 198, 200, 217, 219
complete set, 226, 237, 243
completeness theorem, 7, 103, 104
completion Theorem, 240
compound conditioning, 27
computer program SPIRIT, 173
conditional event, iv-v, 1-2, 11, 15, 18, 21, 25, 28, 30, 37, 40, 42, 49, 51-53, 66, 75, 77-78, 81-82, 84, 93, 101, 110, 113, 120, 153, 160, 162, 171, 184, 189, 191-192, 195-200, 202, 204-205, 212-213, 215-216, 219, 223, 225-231, 233, 236, 247-248, 250-251, 256-259
conditional event algebra, 18, 53, 77, 153, 160, 162, 196-197, 200, 202, 205, 212-213, 216, 231, 233, 248, 251, 256-259
conditional expectation, 76, 78, 85, 87-88, 90
conditional ideal, 100, 104
conditional implication, 108
conditional models, 101, 104, 105
conditional independence, 41
conditional necessity implication, 122
conditional probability, i, iii, v, 2-4, 8, 10, 15-17, 18, 21-26, 28, 30, 31, 45-47, 49, 51, 53, 63, 80-81, 84-85, 109-110, 114-115, 117, 148, 155, 157, 180, 184, 191, 193, 199-200, 206, 208, 222, 225, 228-229, 232, 249, 256
conditional probability logic, v, 8, 10, 45, 47, 109-110, 117, 256
conditional proposition, iii, 2, 8, 10-11, 18, 21-22, 24, 28-29, 33, 46, 49-50, 53, 56-57, 59-60, 66, 69, 75, 78-81, 84, 93, 104-106, 108, 110-113, 117, 119-121, 126, 133, 140-143, 157-159, 170-171, 173, 180, 183, 185-187, 189, 192, 201, 204, 209, 216, 225, 231, 256
conditional random variables, 75, 84-87, 89-90, 93, 180
conjunction property, 129, 131-132, 145-146
conjunctive closure, 121, 131, 136, 137, 140, 147, 149, 154, 156
conjunctive implication, 121, 125, 127
consequent, 54, 65, 200, 205, 215
contrapositive, 10, 22, 42, 107, 108, 126
converse, 22, 42, 108, 114, 123-124, 127, 130, 156, 204
cooperative targeting, 38-41
Copenhagen interpretation, 189, 194, 195

de Broglie, L., 190, 260
De Finetti, Bruno, 2, 24, 46, 47, 54, 79, 230, 256, 258
de Morgan formulas, 202
deduction, v, 25, 41, 64, 82, 119, 120-121, 128-129, 132, 136, 139, 145-146, 148, 162, 176, 191, 193, 200, 218, 226, 230, 236, 247, 256-257, 263
deductive closure, 121
deductive equivalence, 121, 128
deductive extension, 131, 137-139, 142, 145, 153, 160
deductive extension theorem, 137, 138, 139, 145, 153, 160
deductive relation, iii, iv, 18, 22, 33, 119-121, 123, 125, 127-150, 153-154, 156-170, 210, 218, 237, 247-248
deductively closed, iv, 5, 119-121, 129-131, 133, 137, 140, 142, 147, 158, 160, 162-170, 179, 223, 247-248
Delmas-Rigoutsos, Y., 263
dice, 5, 11, 38, 117, 180
Dirac, P., 190, 260

disjunctive implication, 121, 125, 128
disjunctive normal form, 69-70, 73
distributive law, 83, 227, 231
division, iii, 19, 20, 25, 46, 76-78, 81, 85-86, 175, 195
domain, 50, 73, 75-77, 85, 90-91, 110, 194-195, 197-198, 248, 263
Domotor, Zolton, 2, 253
Dubois, Didier, 24, 136, 258, 259
Dummett, Michael, 263

eigenstates, 185, 208
eigenvalues, 184
Einstein, Albert, 189-190, 194, 260-261
elementary deductive relation, 127-128, 130, 132-135, 137, 139, 147, 149, 154, 247
Engesser, K., 248, 263
entanglement, 185, 190, 194, 246, 249, 250
equivalence, 2, 8, 10, 12, 14, 27-28, 48, 50, 53-54, 95-98, 100-101, 106-107, 119-120, 123, 161, 183, 198, 213, 230
equivalence class, 8, 27, 96, 97, 98, 101, 161, 183
equivalence relation, 2, 12, 14, 28, 54, 95-96, 100-101, 119, 183, 198, 230
event fraction, 19, 22, 42
extended definition, 78, 81, 85-87, 90, 93
extended operations, iii, 11, 75-76, 78, 85-86, 113, 176
extended summation, 89
extension, 5-7, 12, 44-45, 69, 72, 100, 102-105, 119-120, 126, 131, 142, 190, 193, 248-250, 256
extension function, 100, 102-104

Feynman, R. P., 193, 261
Fine, T., 190, 260-261
fraction, iii, 19, 25, 173, 175, 193, 201, 225, 250
Fuchs, C. A., 195, 262
fuzzy set, iv, 9, 177-179, 181

Gabbay, D. M., 248, 263
Gabora, L., 263
Gaifman, H., 7, 253
Gleason, A. M., 244, 245, 264
Gödel, Kurt, 7, 103, 254
Goldstein, S., 190, 261
Goodman, Irwin R., 9, 24, 46, 63-64, 73, 197, 254, 256-258, 263
guiding waves, 185
Gupta, Madan M., 9, 254, 256-258

Hailperin, T., 2, 7, 12, 19, 22, 24, 46, 63-64, 254, 258
Halmos, Paul R., 7, 254
Han, T. S., 264
Hardegree, G. M., 199, 263
Heisenberg, W., 190, 246, 249-250, 260
hidden variables, 190, 193, 246, 261
Hilbert space, 183-184, 189, 191, 195-196, 200, 208, 219, 223, 225, 237, 244-245, 251, 263-264
Hintikki, Kaarlo, 254-255
Hoffmann, A., 183, 187, 258, 259
Hooker, J. N., 264

ideal, 4-8, 10, 44, 95-102, 104-105, 107, 109, 111-112, 143, 151, 177-181
idempotency, 113
if - then –, 46
implication, 1-3, 6-7, 64, 107-108, 119, 121-123, 125-126, 128, 148, 154, 157, 176, 184, 223, 229
inapplicability, 83, 125, 140, 147, 159, 176, 185, 186, 190
inapplicable, iii, 23, 29, 31, 32, 49, 51-52, 54-56, 61, 63, 65, 79, 82, 126-127, 148, 158, 186, 189, 192-193, 201, 205, 218, 230-232, 235, 249
independence, 22, 35, 41-42, 66, 76, 90-93, 155, 172
indeterminacy, 185, 189-190, 194, 209, 246, 249-250
indicator function, 47-50, 57, 59, 75, 78-80, 84, 180, 186, 197-198, 204, 225

inference, 64-65, 118-119, 154, 176, 227, 256-258, 260, 264
information theory, 176
inner product, 242, 244
inner-product space, 227
instant, 180, 185, 190
intension, 5-6
interfering waves, 250
intersection, 1, 17-18, 21, 34, 48, 76, 78, 95, 99-100, 111, 118, 130, 134-135, 159, 164, 186, 200
isomorphism, 103-104, 106
iterated conditional, 25-26, 34, 62, 64-65

Jauch, J. M., 228, 262
Jaynes, E. T., 173, 175-176, 189, 264-265

Keisler, H. J., 7, 43, 100, 253
kernel, 7, 44
Khrennikov, A., 195, 262, 264
Kolmogorov, A. N., 1, 13, 21, 46, 105, 109, 181, 195, 229, 254, 261
Koopman, B. O., 193, 261
Kosko, Bart, 9, 254
Krauss, Paul, 2, 7, 255

Landé, A., 204, 262
Laura, R., 183-185, 246, 259, 264, 265
Lehmann, D., 248, 263
Lewis, C. I., 6, 19, 191, 193, 254, 256, 261
Lukasiewicz, J., 9, 95, 254, 265
Lusk, E., 2, 254

material conditional, 2, 12, 15-16, 18, 21, 23, 26, 34, 46, 57, 108, 154, 156, 193, 229, 237
material implication, 2-3
maximum entropy, iv, 153, 155, 172-176, 191, 263-264
maximum entropy distribution, 174-176
maximum information entropy, 156, 172, 176

Mazurkiewicz, Stefan, 7, 24, 46, 254
measurable, 47-49, 57, 59, 78, 105, 180, 197, 204
measurement information, 246
measurement problem, 190-191, 194, 265
Meyer, C. H., 263
modal logic, 8, 255
model, iii, v, 5, 7-8, 12, 21, 43, 45, 101-104, 106, 109-110, 177, 180-181, 248-249, 253
Moore, D., 262
multiplication, 27, 46, 76, 78, 85-86
mutually inconsistent, 7, 183, 190

necessary implication, 122-123
negation, 1, 5, 11-12, 14, 22, 25, 27, 29, 43-44, 46, 48, 51-52, 55-56, 58, 60-62, 66-67, 69, 78-79, 81-82, 86, 124, 155, 184, 192, 196, 198-199, 201-202, 212, 217-218, 220, 223, 230-231
Nguyen, Hung T., 9, 24, 32, 46, 63-64, 254, 256-258, 263
Nilsson, N. J., 46, 256
non-commutative, 226, 235
non-distributive logic, 192
non-falsity implication, 122
non-local, 190, 194
non-locality, 194
non-monotonic, iii, 22, 25, 31, 196, 206, 225, 247-248, 251, 258
Norris, C., 261
Nute, Donald, 2, 255

object language, 12, 114, 130
occurrences, 13, 54, 109, 222
orthoalgebra, 219
orthocomplementation, 196, 210, 220
orthocomplemented lattice, 184, 210, 227, 232
orthogonal, 200, 210-211, 215, 226, 232-234, 237-240, 242-243, 245
orthogonal expansion theorem, 233
orthogonality, iv, 189, 191, 195-196, 210-211, 226, 232, 237

pair-wise disjoint, 13, 239

Papoulis, A., 85, 259
partially defined, 47, 80, 180, 186, 225
partition, 37-38, 239, 241
Pearl, Judea, 35, 258
penguin, 149, 150, 151, 152
Piron, C., 220, 228, 262-263
Pitowsky, Itamar, 244, 262, 264
Planck, Max, 190, 260
Podolsky, B., 261
Pospesel, Howard, 154, 255
Prade, Henri, 9, 24, 136, 255, 258-259
principal DCS, 131, 133-136, 138-139, 141, 145, 147, 150-151, 164
principal deductively closed set, 131, 133
principal ideal, 100, 107
Prior, Arthur N., 46, 185, 255
probabilistically monotonic implication, 121, 125, 128
probability measure, 7, 13, 44, 47-48, 80, 88, 105, 109-110, 178, 180, 186, 244-245
probability space, 13, 15, 47, 49, 66, 84-85, 109, 180, 186
problem, 149, 151
projection operator, 208, 242
Putnam, H., 190-192, 194, 227, 259, 261, 263

quantum measurements, iv, 194-195, 210, 225, 228, 236, 246
quantum mechanics, iv, 184, 189-191, 193-195, 199, 221-223, 226-229, 236-237, 242, 248-251, 260-262, 265
quantum observable, 185, 208, 226, 241
quantum operations, 222, 247
quantum subspace, 223
quantum theory, 4, 189, 260-262, 265
Quine, Willard V., 1, 255, 259

Ramer, Arthur, 183, 187, 258, 259
Rao, C. R., 174, 264
relative complement, 82, 113, 202, 212, 220

relative complementation, 220
Renyi, Alfred, 2, 255
representation theorem, 1, 73, 239-242
Rescher, Nicholas, 8, 192, 255
restricted indicator function, 78, 84
restriction theorem, 240
Rödder, Wilhelm, iv, 155, 173, 191, 257, 261, 263
Rogers, G. S., 24, 46, 256-258
Rosen, N., 261
Rosenbloom, Paul, 7, 69, 255
Russell, Bertrand A. W., 1-2, 183, 229, 255

Sasaki, U., 263
Schay, Geza, 1, 2, 24, 46-47, 63, 196-200, 222-223, 231, 256
Schrödinger, E., 189-190, 246, 249-250, 260, 264
Scott, D., 2, 7, 255
self-referential, 182-183, 248
Shafer, Glen, 6, 255
Shore, J. E., 172, 264
sigma-algebra, 48, 180, 186
simultaneous, iv, 185, 189, 191, 195, 196, 209, 212-213, 215-216, 221, 226-228, 232-233, 235-236, 246, 250
simultaneous falsifiability, 195-196, 221
simultaneous measurements, 209, 227, 246
simultaneous verifiability, iv, 189, 191, 195-196, 212, 221, 226
simultaneously falsifiable, 216, 226, 235
simultaneously verifiable, 184, 189, 212-213, 215-216, 218, 226, 232-237
Smyth, M., 263
Sobocinski, B., 64, 192, 258
standard interpretation, 190
Stone, M. H., 1, 200, 255, 258
subtraction, 76
sum ideal, 5, 95, 101, 120, 247
superposition, 185, 189-191, 195-196, 206-208, 223-234

Suppes, Patrick, 2, 4, 75, 78, 80, 84, 190, 229, 254-255, 258-260, 262-263
surveillance regions, 34-35, 76, 91
symmetric difference, 18
symmetry, 96-97, 103-104, 114, 146, 214-215, 217-218, 234
Tarski, Alfred, 7, 95, 118, 255, 256, 265
three-valued, 9, 58-60, 75, 204-205, 230, 254, 257-258, 261
time, iv-v, 9-11, 20, 38-39, 46, 91, 105, 117, 177, 180-187, 194-195, 197, 205, 225, 241, 245-247, 249, 255, 259, 264
transitivity, 96, 120-121, 124, 132, 137, 142, 147, 171
truth-values, 4, 9, 24, 26, 230
two-valued, 3, 5, 9-10, 15-16, 43, 44, 187, 193
Tyszkiewicz, J., 64, 183, 187, 200, 258-259

undefined, 2-3, 15-16, 27, 49, 51, 54, 57, 61, 63, 66, 75-80, 82, 85, 87, 92, 111, 158, 175, 183, 189, 200, 209
undefined condition, 66, 80
union, 1, 12, 18, 21, 30, 34, 48, 78, 99-100, 111, 138-139, 142, 145-146, 151, 155, 186, 200, 223, 245, 247

value assignments, 185, 245
Vanni, L., 183-185, 246, 259, 264-265
Varadarajan, V. S., 210, 212, 220, 226, 232, 262
variable domain, v, 75, 180, 191, 257
vector, 184, 208, 222-223, 242-245
Venn, John, 29-30, 102, 115, 256
Venn diagram, 29-30, 102, 115
von Neumann, J., 190-191, 227, 260

Walker, E. A., 24, 46, 57, 63, 257, 258
wave-particles, 193
weighted average, 75, 83, 84, 207, 234
Whitehead, Alfred N., 185, 255
wholly true, iii, 178-181, 222
Wilce, A., 190, 191, 260, 262
Winskel, G., 263
work force, 76, 91

Zadeh, Lofti, iv, 9, 177-178, 256, 259
Zanotti, M., 75, 78, 80, 84, 258-260

www.ingramcontent.com/pod-product-compliance
Lightning Source LLC
Chambersburg PA
CBHW071700160426
43195CB00012B/1531